Cargo Work

For Maritime Operations

Seventh Edition

D.J. House

(formerly Kemp & Young's Cargo Work)

AMSTERDAM BOSTON HEIDELBERG LONDON NEW YORK OXFORD
PARIS SAN DIEGO SAN FRANCISCO SINGAPORE SYDNEY TOKYO

Elsevier Butterworth-Heinemann
Linacre House, Jordan Hill, Oxford OX2 8DP
30 Corporate Drive, Burlington, MA 01803

First published as Cargo Work by Stanford Maritime Ltd 1960
Second edition 1965
Third edition 1971
Reprinted 1972, 1974, 1975, 1977
Fourth edition 1980
Fifth edition 1982
Reprinted 1983, 1985, 1987
First published by Butterworth-Heinemann 1990
Reprinted 1991
Sixth edition 1998
Reprinted 2000, 2002, 2003
Seventh edition 2005

British Library Cataloguing in Publication Data
A catalogue record for this book is available from the British Library

Library of Congress Cataloguing in Publication Data
2004118249

ISBN 0 7506 6555 6

For information on all Elsevier Butterworth-Heinemann publications visit our website at www.books.elsevier.com

Typeset by Charon Tec Pvt. Ltd, Chennai, India
www.charontec.com
Printed and bound in Great Britain

Contents

Preface

The world of cargo operations has changed considerably from the days of the open stowage of merchandise. Unitized cargoes in the form of 'containers' or Roll-on, Roll-off cargoes and pallatization have generated a need for alternative handling methods and changing procedures.

The work of the stevedore/longshoreman has moved on to a vastly different role to that previously employed in general cargo holds. The cargo units are labour saving and tend to require a different mode of working. In many cases, ship's crews or rigging gangs have replaced the role of the previous style of dock labour. The fork lift truck and the container gantry have been the source of the major causes of change within the cargo-handling environment and the demise of labour intensive activities.

Unlike the previous editions of '*Cargo Work*', this new text has taken the changes to the industry and included the cargo-handling equipment and the procedures being adopted in our present day. It is anticipated that cargoes can no longer be a stand-alone topic and must incorporate the modern methods of handling, stowage and commodity together.

The two topics of cargoes and handling equipment have therefore been combined in order to appeal to a wider readership and give greater coverage to the prime function of shipping.

This edition has been totally revised by:
D.J. House

Master Mariner
Senior Lecturer Nautical Studies
Marine Author
Patent Holder (GB2240748)

About the author

David House started his sea-going career on general cargo/passenger liners in 1963. During his sea-going career he gained experience of many vessel types and trades, including refrigerated (reefer) vessels to South America on the chilled and frozen meat trade.

His activities included shipping containers from Europe to North America and general cargoes worldwide, during which period he gained extensive knowledge on heavy-lift operations.

His bulk cargo experience was obtained from the carriage of a variety of products, inclusive of grain, sugar, tallow, sulphur and coal.

The types of vessels and various trades in which he was engaged has provided the foundation for this up-to-date version of Kemp & Young's original work.

David House has served on Roll-on, Roll-off vessels, as well as container tonnage, dealing with all aspects of modern cargo-handling techniques: steel cargoes, heavy lifts, special cargoes, foodstuffs, livestock, as well as the bulk commodities and general merchandise. He has been involved as both a Junior and a Senior Cargo Officer, and currently lectures on virtually all nautical subjects at the Fleetwood Nautical Campus.

He has researched and published 13 profusely illustrated Marine publications, which are widely read throughout the maritime world. Amongst his books you can find the following: *Navigation for Masters* (1995); *Marine Survival and Rescue Systems* (1997); *An Introduction to Helicopter Operations at Sea – a Guide for Industry* (1998); *Seamanship Techniques*, Volume III 'The Command Companion' (2000); *Anchor Practice – a Guide for Industry* (2001); *Marine Ferry Transports – an Operators Guide* (2002); *Dry Docking and Shipboard Maintenance* (2003); *Seamanship Techniques*, third edition (2004); *Seamanship Examiner* (2005); *Heavy Lift and Rigging* (in press). www.djhouseonline.com

Acknowledgements

B&V, Industrietechnik GmbH
British Nuclear Fuels
British Standards Institution
Bruntons (Musselburgh) Ltd.
Dubai Dry Docks UAE
International Maritime Organization (publications)
James Fisher Shipping Company
MacGregor International Organisation
Maritime and Coastguard Agency
Motor Ship (published by IPC Industrial Press Ltd.)
Overseas Containers Ltd.
P&O European (Irish Sea) Ferries
Scheuerle Fahrzeugfabrik GmbH
Seaform Design (Isle of Man)
Smit International
TTS – Mongstad AS

Additional photography

Capt. K.B. Millar, Master Mariner, Lect., Nautical Studies of Millar Marine Services
Capt. J.G. Swindlehurst (MN) Master Mariner
Capt. A. Malpass (MN) Master Mariner
Mr M. Gooderman, Master Mariner, B.A. Lecturer Nautical Studies
Mr G. Edwards Ch/Eng (MN) Rtd.
Mr P. Brooks Ch/Off (MN)
Mr J. Leyland (Nautical Lecturer)

I.T. Consultant: Mr C.D. House

List of abbreviations used in the context of Cargo Work

°A	Degrees absolute
AAA	Association of Average Adjusters
ABS	American Bureau of Shipping
AIS	Automatic Identification System
B	Representative of the ship's centre of buoyancy
BACAT	BArge CATamaran
BCH	Bulk Chemical Code
B/L	Bill of Lading
BLU (Code)	The Code of Practice for Loading and Unloading of Bulk Cargoes
BOG	Boil-off gas
BS (i)	Breaking strain
BS (ii)	British Standard
BS (iii)	Broken stowage
BT	Ballast tank
C	Centigrade
CAS	Condition Assessment Scheme
CBM	Conventional buoy mooring
CBT	Clean ballast tank
CCTV	Close Circuit Television
CEU	Car equivalent unit
Ch/Off (C/O)	Chief Officer
cm	Centimetres
CNG	Compressed natural gas
CoF	Certificate of Fitness
C of G	Centre of gravity
COW	Crude oil washing
CO_2	Carbon dioxide

CSO	Company Security Officer
CSS	Cargo Stowage and Securing (IMO Code of Safe Practice of)
CSWP	Code of Safe Working Practice
CTU	Cargo transport unit
cu	Cubic
D	Density
DGN	Dangerous Goods Notice
DNV	Det Norske Veritas
DOC	Document of Compliance
DWA	Dock water allowance
Dwt	Deadweight tonnage
EC	European Community
EDI	Electronic data interchange
EEBDs	Emergency escape breathing devices
EFSWR	Extra flexible steel wire rope
EU	European Union
F (i)	Fresh
F (ii)	Fahrenheit
FloFlo	Float-on, Float-off
FO	Fuel oil
FPSOS	Floating Production Storage Offloading System
FSE	Free surface effect
FSM	Free surface moment
FSRU	Floating storage and re-gasification unit
FSU	Floating storage unit
FSWR	Flexible steel wire rope
ft	Feet
FW	Fresh water
FWA	Fresh water allowance
G	Ship's centre of gravity
G/A	General average
gal	Gallons
GG_1	Representation of the movement of the ship's C of G when moving a weight aboard the vessel.
GM	Metacentric height
grt	Gross registered tonnage
GZ	Ship's righting lever
HCFC	Hydro chlorofluorocarbons
HDFD	Heavy duty, floating derrick
HMSO	Her Majesty's Stationary Office
HP (i)	High pressure
(ii)	Horse power

HSC	High-speed craft
HSE	Health and Safety Executive
HSMS	Hull stress monitoring system
HSSC	Harmonized System of Survey and Certification
IACS	International Association of Classification Societies
IBC	International Bulk Cargo (Code)
ICS	International Chamber of Shipping
IG	Inert gas
IGC	Inert Gas Code
IGS	Inert Gas System
ILO	International Labour Organization
IMDG	International Maritime Dangerous Goods (code)
IMO	International Maritime Organization
IOPP	International Oil Pollution Prevention (certificate)
ISGOTT	International Safety Guide for Oil Tankers and Terminals
ISM	International Safety Management
ISPS	International Ship and Port Facility Security (Code)
ISSC	International Ships Security Certification
ITU	Inter-modal transport unit
K	Representative of the ship's keel
kg (k)	Kilograms (kilo)
KM	Representative of the distance from the ship's keel to the metacentre
kN	Kilo-newtons
kt	Knots
kW	Kilowatt
L	Lumber (loadlines)
LASH	Lighter Aboard SHip (system)
lb	Pounds
LCG	Longitudinal centre of gravity
LEL	Lower explosive limit
LFL	Lower flammable limit
L/H	Lower hold
LNG	Liquefied natural gas
Lo-Lo	Load-on, Load-off
LP	Low Pressure
LPG (i)	Liquid propane gas
(ii)	Liquid petroleum gas
m	Metres
M	Metacentre
MA	Mechanical advantage
MARPOL	Maritime Pollution (convention)

MARVs	Maximum Allowable Relief Value Settings
MCA	Maritime and Coastguard Agency
MCTC (MTC)	Moment to change trim 1 cm
MEPC	Marine Environment Protection Committee
MFAG	Medical First Aid Guide (for use with accidents involving dangerous goods)
MGN	Marine Guidance Notice
MIN	Marine Information Notice
mm	Millimetres
MN	Mercantile Marine (Merchant Navy)
MPCU	Marine Pollution Control Unit
MS	Merchant Shipping Act
MSC (i)	Maritime Safety Committee (of IMO)
MSC (ii)	Mediterranean Shipping Company
MSL	Maximum securing load
MSN	Merchant Shipping Notice
MTSA	Maritime Transport Security Act (US)
MV	Motor vessel
MW	Megawatt
NLS	Noxious liquid substances
NMVOC	Non-methane volatile organic compound
NOS	Not otherwise specified
NPSH	Net positive suction head
NRV	Non-return valve
OBO	Oil, bulk, ore (carrier)
OCIMF	Oil Companies International Marine Forum
ORB	Oil Record Book
P	Port
Pa	Pascal
P/A System	Public Address System
PCC	Pure car carrier
PCTC	Pure car and truck carrier
PEL	Permissible exposure limit
PFSP	Port Facility Security Plan
P/L	Protective location
PMA	Permanent means of access
PNG	Pressurized natural gas
ppm	Parts per million
PSC	Port State Control
psi	Pounds per square inch
PSO	Port Security Officer
P/V	Pressure vacuum

R	Resistance
RD	Relative density
RMC	Refrigerated Machinery Certificate
Ro-Pax	Roll-on, Roll-off plus Passengers
Ro-Ro	Roll-on, Roll-off
rpm	Revolutions per minute
RVP	Reid vapour pressure
S (Stbd)(i)	Starboard
S (ii)	Summer
SBM	Single buoy mooring
SBT	Segregated ballast tank
SCBA	Self-contained breathing apparatus
SeaBee	Sea barge
SECU	StoraEnso Cargo Unit
SF	Stowage factor
S.I.	Statutory Instrument
SMC	Safety Management Certificate
SOLAS	Safety of Life at Sea (Convention)
SOPEP	Ships Oil Pollution Emergency Plan
SO_x	Oxides of sulphur
SPG	Self-supporting Prismatic-shape Gas tank
SRV system	Shuttle and Re-gasification Vessel system
SSO	Ship Security Officer
SSP	Ship Security Plan
SW	Salt water
SWL	Safe working load
SWR	Steel wire rope
T	Tropical
T/D	Tween deck
TEU	Twenty feet equivalent unit
TF	Tropical fresh
Tk	Tank
TLVs	Threshold limit values
TPC	Tonnes per centimetre
TWA	Time weighted average
U	Union Purchase – safe working load
UEL	Upper explosive limit
UFL	Upper flammable limit
UHP	Ultra high pressure
UK	United Kingdom
UKC	Under keel clearance
ULCC	Ultra large crude carrier
ULLNGC	Ultra large liquefied natural gas carrier

UN	United Nations
US	United States
USA	United States of America
USCG	United States Coast Guard
U-SWL	Union Rig – safe working load
VCM	Vinyl chloride monomer
VDR	Voyage Data Recorder
VLCC	Very large crude carrier
VOCs	Volatile organic compounds
VR	Velocity ratio
W (i)	Winter
W (ii)	Representative of the ship's displacement
WBT	Water ballast tank
WC	Water-closet (Toilet)
W/L	Waterline
WNA	Winter North Atlantic
wps	Wires per strand
YAR	York Antwerp Rules (2004)

Conversion and measurement table

Imperial/metric measurement

1 in. = 2.5400 cm　　1 cm = 0.3937 in.
1 ft = 0.3048 m　　1 m = 3.2808 ft

1 in.^{2} = 6.4516 cm^2　　1 cm^2 = 0.1550 in.^{2}
1ft^2 = 0.09293 m^2　　1 m^2 = 10.7639 ft^2

1 in.^{2} = 16.3871 cm^3　　1 cm^3 = 0.0610 m^3
1 ft^3 = 0.02832 m^3　　1 m^3 = 35.3146 ft^2
(where in. represents inches)

Metres to feet							
Cm	Feet	Metres	Feet	Metres	Feet	Metres	Feet
1	0.03	1	3.28	17	55.77	60	196.85
2	0.06	2	6.56	18	59.06	70	229.66
3	0.09	3	9.84	19	62.34	80	262.47
4	0.13	4	13.12	20	65.62	90	295.28
5	0.16	5	16.40	21	68.90	100	328.08
6	0.19	6	19.69	22	72.18	200	656.17
7	0.22	7	22.97	23	75.46	300	984.25
8	0.26	8	26.25	24	78.74	400	1312.33
9	0.30	9	29.53	25	82.02	500	1640.42
10	0.33	10	32.81	26	85.30	600	1968.50
20	0.66	11	36.09	27	88.58	700	2296.58
30	0.98	12	39.37	28	91.86	800	2624.66
40	1.31	13	42.65	29	95.15	900	2952.74
50	1.64	14	45.93	30	98.43	1000	3280.83
60	1.97	15	49.21	40	131.23		
70	2.30	16	52.49	50	164.04		
80	2.62						
90	2.95						

Feet to metres							
Inches	Metres	Feet	Metres	Feet	Metres	Feet	Metres
1	0.03	1	0.30	80	24.38	800	243.84
2	0.05	2	0.61	90	27.43	850	259.08
3	0.08	3	0.91	100	30.48	900	274.32
4	0.10	4	1.22	150	45.72	950	289.56
5	0.13	5	1.52	200	60.96	1000	304.80
6	0.15	6	1.83	250	76.20	1100	335.28
7	0.18	7	2.13	300	91.44	1200	365.76
8	0.20	8	2.44	350	106.68	1300	396.24
9	0.23	9	2.74	400	121.92	1400	426.72
10	0.25	10	3.05	450	137.16	1500	457.20
11	0.28	20	6.10	500	152.40	2000	609.60
12	0.30	30	9.14	550	167.64	3000	914.40
		40	12.19	600	182.88	4000	1219.20
		50	15.24	650	198.12	5000	1524.00
		60	18.29	700	213.36		
		70	21.34	750	228.60		

Tonnage and fluid measurement

	US gallons	Imperial gallons	Capacity cubic feet
1 gal (imp)	×1.2	×1	×0.1604
1 gal (US)	×1.0	×0.8333	×0.1337
1 ft^3	×7.48	×0.2344	×1.0
1 l	×0.2642	×0.22	×0.0353
1-tonne fresh water	×269	×224	×35.84
1-tonne salt water	×262.418	×218.536	×35

Weight	Short ton	Long ton	Metric tonne
Long ton (imp)	×1.12	×1.0	×1.01605
Short ton (USA)	×1.0	×0.89286	×0.90718
Metric tonne	×1.10231	×0.98421	×1.0

Grain	Bushel (imp)	Bushel (USA)	Cubic feet
1 Bushel (imp)	×1.0	×1.0316	×1.2837
1 Bushel (USA)	×0.9694	×1.0	×1.2445
1 ft^3	×0.789	×0.8035	×1.0

Miscellaneous

1 lb = 0.45359 kg   1 kg = 2.20462 lb
1 ft^3/tonne = 0.16 imp gal/tonne
1 tonne/m^3 = 0.02787 tonne/ft^3
1 m^3/tonne = 35.8816 ft^3/tonne

Chapter 1

General principles of the handling, stowage and carriage of cargoes

Introduction

The transport of cargoes dates back through the centuries to the Egyptians, the Phoenicians, ancient Greeks and early Chinese, long before the Europeans, ventured beyond the shores of the Atlantic. Strong evidence exists that the Chinese Treasure Ships traded for spices, and charted the Americas, Antarctica, Australia and the Pacific and Indian Oceans, before Columbus reportedly discovered America.*

The stones for the Pyramids of Egypt had to be brought up the River Nile or across the Mediterranean and this would reflect the means of lifting heavy weights, and transporting the same was a known science even before the birth of Christ. Marco Polo reported 200 000 vessels a year were plying the Yangtze River of China in 1271 and it must be assumed that commerce was very much alive with a variety of merchandise being transported over water.

Products from the world's markets have grown considerably alongside technology.

Bigger and better ships feed the world populations and the methods of faster and safer transport have evolved over the centuries.

The various cargoes and merchandise may be broadly divided into the following six types:

1. Bulk solids
2. Bulk liquids
3. Containerized units
4. Refrigerated/chilled
5. General, which includes virtually everything not in (1), (2), (3) and (4) above
6. Roll-on, Roll-off (Ro-Ro) cargoes.

---

*Menzies, G. (2002) 1421 The Year China Discovered the World, Bantam Press.

Bulk cargoes can be loaded and discharged from a ship quickly and efficiently. Conversely, we have yet to see 10 000 tonnes of grain being loaded into a Jumbo Jet. Ships remain the most efficient means of transport for all cargo parcels of any respectable weight or size.

It is here that the business of how it is loaded, how it is stowed and subsequently shipped to its destination is investigated. Later chapters will deal with specifics on the commodities, but the methods of handling prior to starting the voyage and the practical stowage of goods, should be considered an essential element of the foundation to successful trade.

Definitions and cargo terminology

Air draught – means the vertical distance from the surface of the water to the highest point of the ship's mast or aerial.

Bale space capacity – is that cubic capacity of a cargo space when the breadth is measured from the inside of the cargo battens (spar ceiling) and the measured depth is from the wood tank top ceiling to the underside of the deck beams. The length is measured from the inside of the fore and aft bulkhead stiffeners.

Broken stowage – is defined as that space between packages which remains unfilled. The percentage that has to be allowed varies with the type of cargo and with the shape of the ship's hold. It is greatest when large cases are stowed in an end hold or at the turn of a bilge.

Cargo information – means appropriate information relevant to the cargo and its stowage and securing which should specify, in particular, the precautions necessary for the safe carriage of that cargo by sea.

Cargo plan – a ship's plan which shows the distribution of all cargo parcels stowed on board the vessel for the voyage. Each entry onto the plan would detail the quantity, the weight and the port of discharge. The plan is constructed by the Ship's Cargo Officer and would effectively show special loads such as heavy-lifts, hazardous cargoes, and valuable cargo, in addition to all other commodities being shipped.

Cargo runner – a general term used to describe the cargo lifting wire used on a derrick. It may be found rove as a 'single whip' or doubled up into a 'gun tackle' (two single blocks) or set into a multi-sheave lifting purchase. It is part of the derricks 'running rigging' passing over at least two sheaves set in the head block and the heel block, prior to being led to the barrel of the winch. Normal size is usually 24 mm and its construction is flexible steel wire rope (FSWR) of 6 × 24 wires per strand (wps).

Cargo securing manual – a manual that is pertinent to an individual ship, and which will show the lashing points and details of the securing of relevant cargoes carried by the vessel. It is a ship's reference which specifies the on-board securing arrangements for cargo units, including vehicles and containers, and other entities. The securing examples are based on the transverse, longitudinal and vertical forces which may arise during adverse weather conditions at sea. The manual is drawn up to the standard contained in Maritime Safety Committee (MSC) Circular of the Organization, MSC/Circ. 745.

Cargo ship – defined as any ship which is not a 'Passenger Ship', troop ship, pleasure vessel or fishing boat.

Cargo spaces – (e.g. cargo hold) – means all enclosed spaces which are appropriate for the transport of cargo to be discharged from the ship. Space available for cargo may be expressed by either the vessel's deadweight or her cubic capacity in either bale or grain space terms.

Cargo unit – includes a cargo transport unit and means wheeled cargo, vehicles, containers, flat pallet, portable tank packaged unit or any other cargo and loading equipment or any part thereof, which belongs to the ship and which is not fixed to the ship.

Centre of buoyancy – is defined as the centre of the underwater volume; that point through which all the forces due to buoyancy are considered to act.

Centre of gravity (C of G) – is defined as that point through which all the forces due to gravity are considered to act. Each cargo load will have its own C of G.

Dangerous goods – are defined as such in the Merchant Shipping (Dangerous Goods and Marine Pollutants) Regulations 1990.

Deadweight – means the difference in tonnes between the displacement of a ship at the summer load waterline in water of specific gravity of 1025, and the lightweight of the ship.

Deadweight cargo – is cargo on which freight is usually charged on its weight. While no hard and fast rules are in force, cargo stowing at less than 1.2 m^3/tonne (40 ft^3/tonne) is likely to be rated as deadweight cargo.

Dunnage – an expression used to describe timber boards which can be laid singularly or in double pattern under cargo parcels to keep the surface of the cargo off the steel deck plate. Its purpose is to provide air space around the cargo and so prevent 'cargo sweat'. Heavy-lift cargoes would normally employ heavy timber bearers to spread the load and dunnage would normally be used for lighter-load cargoes.

Flemish Eye – a name given to a Reduced Eye made of three strands (not six), spliced into the end of a cargo runner which is secured to the barrel of a winch (alternative names are Spanish Eye, or Reduced Eye).

Flemish hook – a large hook, often used in conjunction with the lower purchase block in the rigging of a heavy-lift derrick. The hook can be opened to accommodate the load slings and then bolt locked.

Floodable length – the maximum length of a compartment that can be flooded to bring a damaged vessel to float at a waterline which is tangential to the margin line. *Note*: In determining this length account must be taken of the permeability of the compartment.

Freight – the term used to express the monetary charge which is levied for the carriage of the cargo.

Gooseneck – the bearing and swivel fitment, found at the heel of a derrick which allows the derrick to slew from port to starboard, and luff up and down when in operation.

Grain capacity – is that cubic capacity of a cargo space when the length, breadth and depth are measured from the inside of the ship's shell plating, all allowances being made for the volume occupied by frames and beams.

Gross tonnage – is defined by the measurement of the total internal capacity of the ship. GT being determined by the formula: GT = KiV where

$$Ki = 0.2 + 0.02 \text{ Log } 10V$$

$$V = \text{Total volume of all enclosed spaces in cubic metres}$$

Hallen universal swinging derrick – a single swinging derrick with a lifting capacity of up to about 100 tonnes safe working load (SWL) The original design employed a 'D' frame, to segregate the leads of the combined slewing and topping lift guys. The more modern design incorporates 'outriggers' for the same purpose.

Hounds Band – a lugged steel band that straps around a 'mast'. It is used to shackle on shrouds and stays. It is also employed to secure 'Preventor Backstays' when a heavy derrick is being deployed in order to provide additional strength to the mast structure when making the heavy lift.

Load density plan – a ships plan which indicates the deck load capacity of cargo space areas of the ship. The Ship's Chief Officer would consult this plan to ensure that the space is not being overloaded by very dense, heavy cargoes.

Long tonne – a unit of mass weight, equal to 2240 lb (tonne).

Luffing – a term which denotes the movement of a crane jib or derrick boom to move up or down, i.e. 'luff up' or 'luff down'.

Luffing derrick – a conventional single swinging derrick rigged in such a manner that permits the derrick head to be raised and lowered to establish any line of plumb, as opposed to static rigged derricks, as with a 'Union Purchase Rig'.

Measurement cargo – is cargo on which freight is usually charged on the volume occupied by the cargo. Such cargo is usually light and bulky stowing at more than 1.2 m^3 per tonne (40 cu. ft./tonne), but may also be heavy castings of an awkward shape where a lot of space is occupied.

Passenger Ship – a ship designed to carry more than 12 passengers.

Permeability – in relation to a compartment space means the percentage of that space which lies below the margin line which can be occupied by water. *Note*: various formulae within the Ship Construction Regulations are used to determine the permeability of a particular compartment. Example values are spaces occupied by cargo or stores 60%, spaces employed for machinery 85%, passengers and crew spaces 95%.

Permissible length – of a compartment having its centre at any point in the ships length is determined by the product of the floodable length at that point and the factor of subdivision of the vessel:

$$\text{permissible length} = \text{floodable length} \times \text{factor of subdivision.}$$

Riding turn – an expression that describes a cross turn of wire around a barrel of a winch, or stag horn. It is highly undesirable and could cause the load to jump or slip when in movement. The condition should be cleared as soon as possible.

Ring bolt – a deck ring or 'pad eye' often used in conjunction with a doubling plate or screw securing. It is employed to provide an anchor point for associated rigging around a derrick position.

Running rigging – a descriptive term used to describe wire or cordage ropes which pass around the sheave of a block (see also 'Standing Rigging'). Where steel wire ropes are employed for running rigging they are of a flexible construction, examples include: 6×24 wps and 6×36 wps.

Safe working load – an acceptable working tonnage used for a weight-bearing item of equipment. The marine industry uses a factor of one-sixth the breaking strain (BS) to establish the safe working value.

Safety tongue – a spring clip sealing device to cover the jaw of a lifting hook. It should be noted that these devices are not fool proof and have been known to slip themselves unintentionally. The tongue is meant to replace the need of 'mousing' the hook, and is designed to serve the same purpose as a 'mousing'.

Schooner guy – a bracing guy which joins the spider bands at the derrick heads of a 'Union Purchase Rig'.

Sheer legs – a large lifting device employed extensively within the marine industry. It is constructed with a pair of inclined struts resembling a crane, although the action when working is similar to a craning activity. (Smaller versions of sheer legs were previously used within the marine industry on tankers to hoist pipelines on board or more commonly found in training establishments for training cadets in rigging applications.) The modern day sheer legs are now found on floating heavy-lift (crane) barges and employed for extreme lifting operations usually with 'project cargoes'.

Shore – a term used to describe a support, given to decks, bulkheads or cargo. They are usually timber, but may be in the form of a metal stanchion, depending on the intended use (see tomming).

Slings – a term which describes the lifting strops to secure the load to be hoisted to the lift hook of the derrick or crane. Slings may be manufactured in steel wire rope, chains, rope or canvas.

Snatch block – a single sheave block, often employed to change the direction of lead, of a wire or rope. The block has a hinged clamp situated over the 'swallow' which allows the bight of a wire or rope to be set into the block without having to pull the end through.

Snotter – a length of steel wire with an eye in each end. Employed around loads as a lifting sling, with one eye passed through the other to tighten the wire around the load.

Speed crane – modern derrick design with multi-gear operation which operates on the principle of the single jib, point loading crane.

Spider band – a steel lugged strap found around the head of a derrick which the rigging, such as the topping lift and guys are shackled onto. The equivalent on a mast structure is known as a 'Hounds Band'.

Spreader – a steel or wood batten which effectively spreads the wire sling arrangement wider apart when lifting a large area load. Use of such a spreader generally provides greater stability to the movement of the weight. Formerly referred to as a lifting beam.

Stabilizers – Steel outriders, often telescopic in design and fitted with spread feet, which are extended from the base unit of a shoreside mobile crane. Prior to taking the load the stabilizers are set to ensure that the load on the crane jib will not cause the crane to topple. (Not to be confused with ship stabilizers fitted to ships to reduce rolling actions of the vessel when at sea.)

Standing Rigging – a term used to describe fixed steel wire rope supports. Examples can be found in ship's stays and shrouds. Construction of Standing Rigging is usually 6×6 wps.

Stowage factor – this is defined as that volume occupied by unit weight of cargo. Usually expressed as cubic metres per tonne (m^3/tonnes) or cubic feet per tonne (ft^3/tonne). It does not take account of any space which may be lost due to 'broken stowage'. A representative list of stowage factors is provided at the end of this book.

Subdivision factor – the factor of subdivision varies inversely with the ship's length, the number of passengers and the proportion of the underwater space used for passenger/crew and machinery space. In effect it is the factor of safety allowed in determining the maximum space of transverse watertight bulkheads, i.e. the permissible length.

Tomming off – an expression that describes the securing of cargo parcels by means of baulks of timber. These being secured against the cargo to prevent its movement if and when the vessel is in a seaway and experiencing heavy rolling or pitching motions (alternative term is 'shore').

Tonne – originated from the word 'tun' which was a term used to describe a wine cask or wine container, the capacity of which was stated as being 252 gallons as required by an Act of 1423, made by the English Parliament. It is synonymous that 252 gallons of wine equated to approximately 2240 lb, '1 tonne' as we know it today.

Trunnion – a similar arrangement to the 'gooseneck' of a small derrick. The Trunnion is normally found on intermediate size derricks of 40 tonnes or over. They are usually manufactured in cast steel and allow freedom of movement from the lower heel position of the derrick.

Tumbler – a securing swivel connection found attached to the 'Samson Post' or 'Mast Table' to support the topping lift blocks of the span tackle.

'U' bolt – a bolt application which secures the reduced eye of a cargo runner to the barrel of a winch.

Union Plate – a triangular steel plate set with three eyelets used in 'Union Rig' to join the cargo runners and hook arrangement when a 'triple swivel hook' is not employed. It can also be used with a single span, topping lift derrick to couple the downhaul with the chain preventor and bull wire. Sometimes referred to as 'Monkey Face Plate'.

Union Rig – Alt; Union Purchase Rig. A derrick rig which joins two single swinging derricks to work in 'Union' with cargo runners joined to a triple swivel hook arrangement known as a 'Seattle Hook' or 'Union Hook'. The rig was previously known as 'Yard and Stay' and is a fast method of loading/discharging lighter parcels of cargo. Union Rig operates at approximately one-third of the SWL of the smallest derrick of the pair.

Velle Derrick – a moderate heavy-lift derrick that can be operated as a crane by a single operator. The derrick is constructed with a 'T' bridle piece at the head of the derrick which allows topping lift wires to be secured to act in way of slewing guys and/or topping lift.

Walk back – an expression which signifies reversing the direction of a winch in order to allow the load to descend or the weight to come off the hoist wires.

Weather deck – means the uppermost complete deck exposed to the weather and the sea.

Wires per strand – an expression (abbreviated as wps) which describes the type of construction of the strands of a steel wire rope.

Yard and Stay – alternative descriptive term for Union Purchase Rig.

Conventional general cargo handling

Cargo gear

Derricks, cranes and winches, together with their associated fittings should be regularly overhauled and inspected under a planned maintenance schedule, appropriate to the ship. Winch guards should always be in place throughout winching operations and operators should conform to the Code of Safe Working Practice (CSWP) (Figure 1.1).

Only certificated tested wires, blocks and shackles should be used for cargo handling and lifting operations.

> ***Note:*** *Wire ropes which have broken wires in strands should be replaced. Whenever 10% of wires are broken in any eight (8) diameters length, the wire should be condemned. Guy pennants, blocks and tackles should be kept in good condition.*

Fig. 1.1 The conventional 'general cargo' vessel 'Sunny Jane' lies port side to, alongside in the Port of Amsterdam. The vessel is fitted with conventional derricks, supported by bi-pod mast structures.

Derrick rigs – Union Purchase Method

The Union Purchase Method of rigging derricks is perhaps the most common with conventional derrick rigs (Figure 1.2). With this operation, one of two derricks plumbs the hatch and the other derrick plumbs overside. The two runner falls of the two derricks are joined together at the cargo 'Union Hook' (this is a triple swivel hook arrangement sometimes referred to as a 'Seattle Hook'). The load is lifted by the fall which plumbs the load, when the load has been lifted above the height of the bulwark or ship's rail, or hatch coaming, the load is gradually transferred to the fall from the second derrick (Figure 1.3).

Cargo movement is achieved by heaving on one derrick runner and slacking on the other. The safe working angle between the runners is 90° and should never be allowed to exceed 120°. There is a danger from overstressing the gear if unskilled winch drivers are employed or if winch drivers do not have an unobstructed view of the lifting/lowering operation. In the latter case, signallers and hatch foremen should always be employed within line of sight of winch operators.

The CSWP for Merchant Seaman provides a code of hand signals for use in such cargo operations.

Single swinging derricks

The conventional derrick was initially evolved as a single hoist operation for the loading and discharging of weights. It was the basic concept as an aid which became popular when combined within a 'Union Rig'. However,

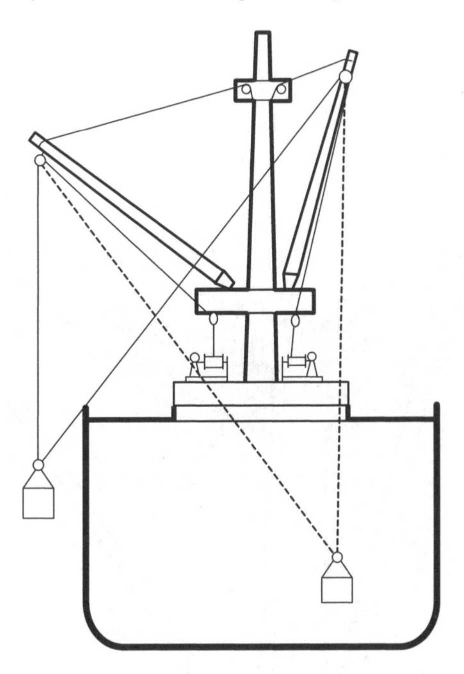

Fig. 1.2 Union purchase. Derrick rig.

Fig. 1.3 The conventional derrick rig. Modern general cargo vessel rigged with conventional 5 tonne SWL derricks and steel hatch covers. The derricks can be rigged to operate as single swinging derricks or rigged in 'Union Purchase' SWL (U) = 1.6 tonnes. Such vessels are in decline because of the growth in unit load 'Container and Ro-Ro, Traffic'.

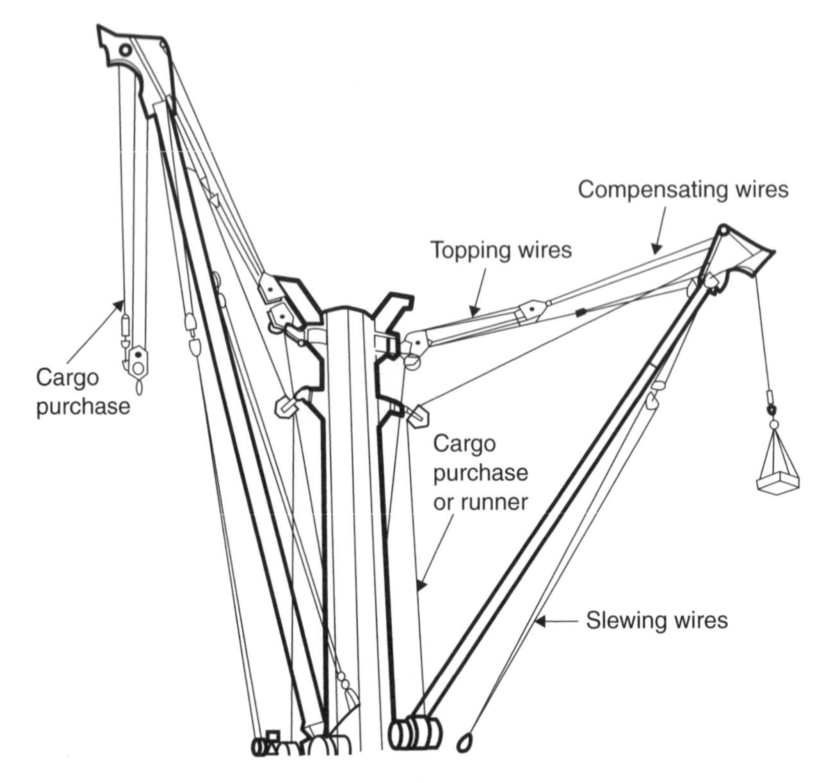

Fig. 1.4 Single swinging derrick.

improved materials and better designs have created sophisticated, single derricks in the form of the 'Hallen', the 'Velle' and the more popular speed cranes. All of which now dominate the reduced activities of general cargo ships (Figure 1.4).

Where the single swinging derrick concept has been retained is in the arena of the heavy-lift operation. Here conventional 'Jumbo Derricks', of the single swinging variety, are still employed amongst specialist rigs as 'Stuelckens' and heavy-lift ships.

Specialized derrick rigs

The many changes which have occurred in cargo-handling methods have brought about extensive developments in specialized lifting gear. These developments have aimed at efficient and cost-effective cargo handling and modern vessels will be equipped with some type of specialist rig for operation within the medium to heavy-lift range.

The 'Hallen derrick'

This is a single swinging derrick which is fast in operation and can work against a list of up to 15°. They are usually manufactured in the 25–40

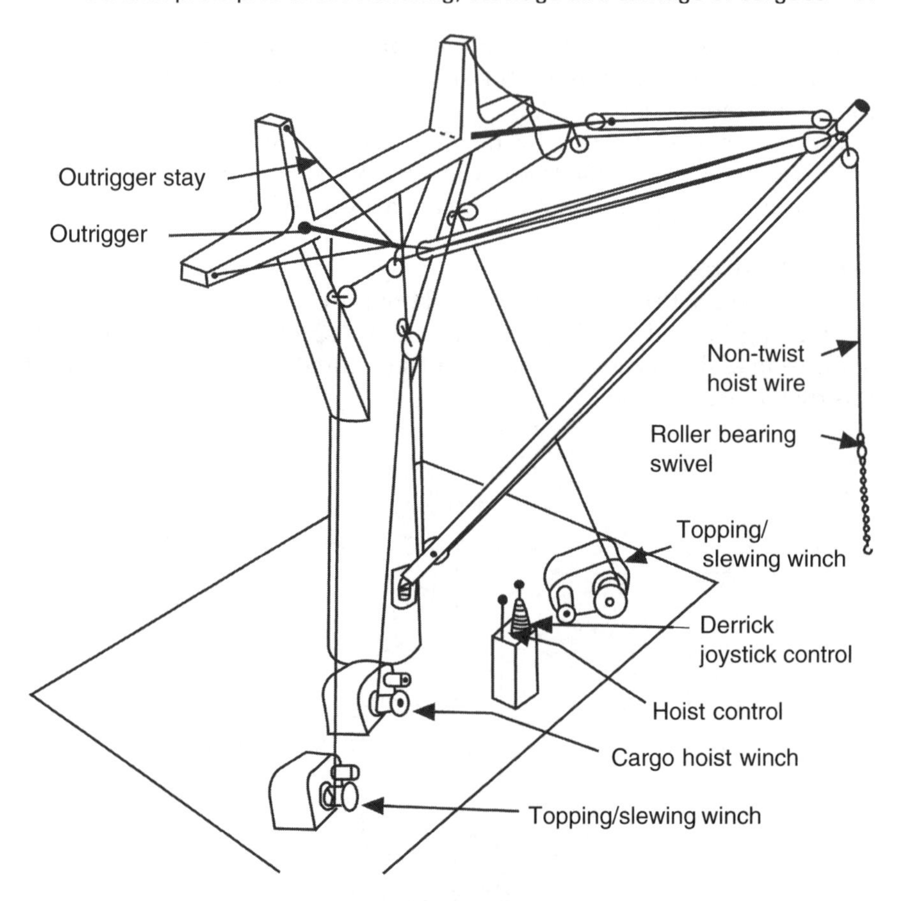

Fig. 1.5 Hallen Derrick.

tonne SWL range and, when engaged, operate under a single-man control (Figure 1.5).

Joystick control for luffing and slewing is achieved by the Port and Starboard slewing guys being incorporated into the topping lift arrangement. Use of the outriggers from a 'Y' mast structure provides clear leads even when the derrick is working at 90° to the ships fore and aft line. A second hoist control can be operated simultaneously with the derrick movement.

As a one-man operation, it is labour saving over and above the use of conventional derricks, while at the same time keeps the deck area clear of guy ropes and preventors. Should heavy loads be involved only the cargo hoist would need to be changed to satisfy different load requirements.

The 'Hallen Derrick' has a similar concept to the 'Velle', in that the topping lift arrangement and the slewing wires are incorporated together and secured aloft, clear of the lower deck. The outreach and slew are wide achieved by the 'T' yoke on the Velle Derrick and by outriggers with the Hallen.

Both systems are labour saving and can be operated by a single controller, operating the luffing and slewing movement together with the cargo hoist movement.

The Hallen is distinctive by the 'Y' mast structure that provides the anchor points for the wide leads. The derrick also accommodates a centre lead sheave to direct the hoist wire to the relevant winch.

'Velle Derrick'

Similar in design to the 'Hallen' but without use of outriggers. The leads for the topping lift and slewing arrangement are spread by a cross 'T' piece at the head of the derrick. A widespread structured mast is also a feature of this rig (Figures 1.6 and 1.7).

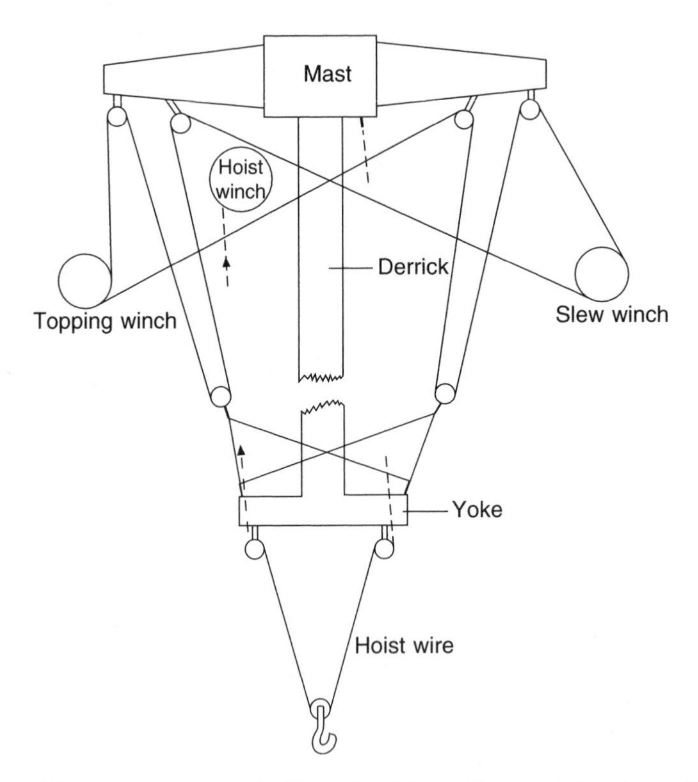

Fig. 1.6 Rigging system on the Velle Derrick. Luffing and slewing actions of the derrick are powered by two winches each equipped with divided barrels to which the bare ends of the fall wires are secured.

Again it is a single-man operation, with clear decks being achieved while in operation. Generally, the 'Velle' is manufactured as a heavier rig and variations of the design with a pivot cross piece at the derrick head are used with multi-sheave purchases to accept the heavy type load.

Fig. 1.7 Velle derrick.

Table 1.1 SWLs for cordage and FSWR

	Material	*Structure*	*BS formula*	*SWL @ one-sixth BS*
Cordage	Manila	3 stranded hawser laid	$2D^2/300$	$2D^2/1800$
	Polypropylene	3 stranded hawser laid	$3D^2/300$	$3D^2/1800$
	Terylene	3 stranded hawser laid	$4D^2/300$	$4D^2/1800$
	Nylon	3 stranded hawser laid	$5D^2/300$	$5D^2/1800$
FSWR	FSWR	6×24 wps	$20D^2/500$	$20D^2/3000$
	Grade 1, stud chain	12.5–120 mm	$20D^2/600$	$20D^2/3600$

Working with a lifting plant

At no time should any attempt be made to lift weights in excess of the SWL of the weakest part of the gear. The SWL is stamped on all derricks, blocks and shackles as well as noted on the 'test certificates'. Wire ropes are delivered with a test certificate on which will be found the SWL of the wire.

Assuming that the SWL is one-sixth of the BS, the regulations require a minimum of one-fifth. The approximate SWL of various materials can be obtained from the formula shown in Table 1.1.

When lifting loads in excess of about 1.5 tonnes, steam winches should generally be used in double gear. Electric winches are usually fused for a SWL of up to about 3 tonnes. For loads in excess of 2–3 tonnes it would be normal practice with conventional derricks to double up the rig, as opposed to operating on a single part runner wire.

Derricks may be encountered with two SWL marks on them. In such cases the lesser value is usually marked with a 'U' signifying the SWL for use in Union Purchase Rig. In the event the derrick is not marked, and intended for use in a Union Rig, the SWL is recommended not to exceed one-third of the smallest of the two derricks (approx).

Use of lifting purchases

The purchase diagrams shown are rigged to disadvantage. The velocity ratio (VR) is increased by '1' if the tackle is rigged to advantage.

The required purchase (the common ones are illustrated in Figure 1.8). The stress factors incurred with their use can be found by the following formula assuming 10% for friction:

$$S \times P = W + nW\ (10/100)$$

where S is the stress in the hauling part; P is the power gained by the purchase (this is the same as the number of rope parts at the moving block); n is the number of sheaves in the purchase; W is the weight being lifted 10, which is the numerator of the fraction, is an arbitrary 10% allowance for friction.

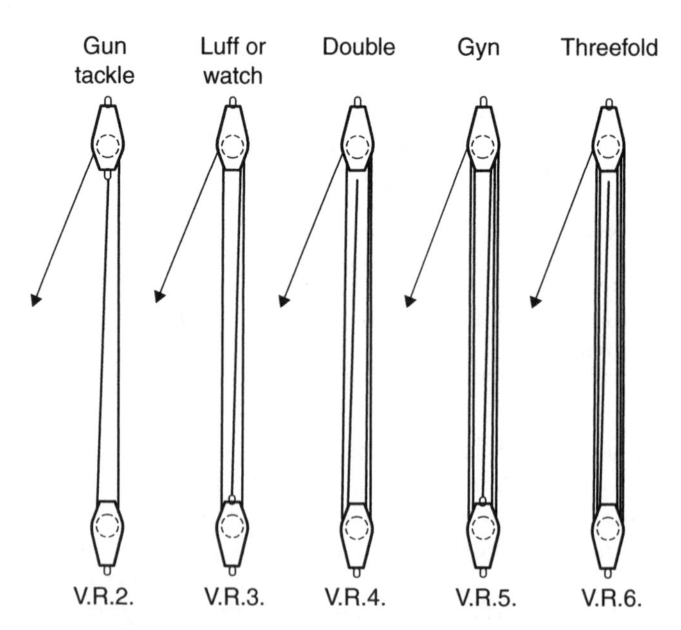

Fig. 1.8 All tackles rove to disadvantage and VRs stated for this rig (when tackles are rove to advantage add + 1 to the VR).

Cargo-handling equipment – condition and performance

Before any cargo operation takes place it is essential that the Chief Officer is confident that the ships lifting equipment and associated loading/discharge facilities are 100% operational and free of any defects. Under the Lifting Plant Regulations, the International Safety Management (ISM) Code, and ship's planned maintenance schedule all-cargo-handling equipment could expect to be inspected and maintained at regular intervals.

In the case of lifting plant, derricks, cranes, shackles, wires, etc. the following test times would be required:

1. after installation when new
2. following any major repair
3. at intervals of every 5 years.

Testing and inspection of plant

Cargo lifting appliances must be inspected to establish that they are correctly rigged on every occasion they are used. To this end, the Chief Officers would normally delegate this duty to the Deck Cargo Officers to check the rig prior to commencing loading or discharge operations.

A thorough inspection would also take place annually by a 'competent person', namely the Chief Officer himself. This duty would not be delegated to a Junior Officer. This inspection would cause a detailed inspection to take place of all aspects – hydraulic, mechanical and electrical – of the lifting appliances. All wires would be visually inspected for defects and the mousing on shackles would be sighted to be satisfactory. The 'gooseneck' of derricks and all blocks would be stripped down and overhauled.

Thorough inspections would detect corrosion, damage, hairline cracks and excessive wear and tear. Once defects are found corrective action would be taken to ensure that the plant is retained at 100% efficiency. These inspections would normally be carried out systematically under the ship's planned maintenance schedule. This allows a permanent record to be maintained and is evidence to present to an ISM Auditor.

Testing plant

Lifting appliances are tested by a cargo surveyor at intervals of 5 years, or following installation or repairs. The test could be conducted by either of two methods:

1. By lifting the proof load, and swinging the load through the derrick or crane's operating arc, as per the ship's rigging plan. This test is known as the 'dynamic test' and concrete blocks of the correct weight are normally used to conduct this operation.
2. The static test is carried out employing a 'dynamometer' secured to the lifting point of the rig and an anchored position on the deck. The proof load weight is then placed on the rig and measured by the dynamometer, to the satisfaction of the surveyor.

Certification

Once the testing has been completed satisfactorily, each lifting apparatus would be issued with a test certificate and the Chief Officer would retain all certificates in the 'Register of Ships Lifting Appliances and Cargo-Handling Gear'.

In addition to these test certificates all shackles, wires, blocks, etc. would be purchased as proof tested and delivered to the vessel with its respective certificate. These would be retained in the Chief Officers Register. The SWL and the certificate number are found stamped into the binding straps of each block. Grease recesses are found inside the bush and inside the inner-bearing surface of the centre of each sheave. The 'axle bolt' is of a square cross-section to hold the bearing 'bush', this allows the sheave to rotate about the bush. In the event that a shackle or block is changed, the certificate in the register would also be changed, so keeping the ships records up to date (Figure 1.9).

Derrick maintenance

As with many items of equipment, derrick rigs must similarly be checked and seen to be correctly rigged on every occasion prior to their engagement. It would be normal practice for the Ship's Chief Officer to delegate this supervisory task to the duty Deck/Cargo Officer before loading or discharge operations is allowed to commence.

In addition to the regular working checks, all lifting gear should undergo an annual inspection by a responsible person, namely the Ship's Chief Officer. This annual inspection is never delegated but would be carried out under the scrutiny of the ship's mate. The annual inspection would entail the overhaul and total inspection of all the derrick's moving parts inclusive of the head and heel blocks, the lifting purchase blocks, the topping lift and runner wires. The condition of the guys would also be inspected and the emphasis would be placed on the main weight-bearing element of the 'gooseneck'.

The annual inspections do not usually require the derrick to be tested unless a degrading fault is found in the rig, necessitating a new part or a replacement part to be used. Testing normally taking place at 5-yearly intervals or if repairs have been necessary or in the event of the derricks being brought back into use, after a period of lay up. If testing is required, this would be carried out in the presence of a cargo surveyor and the lifting gear would have to show handling capability up to the proof load.

In order to conduct an annual inspection, the Chief Officer would order the complete overhaul of all the blocks associated with the derrick rig. Normal practice would dictate that the ship's boatswain would strip the blocks down and clean off any old grease and clear the grease recesses in the bush and the inside of the sheaves. The 'bolt' would be extracted and the bush bearing would be withdrawn. Inspection by the Senior Officer would take place and any signs of corrosion, hairline cracks or excessive wear and tear would be monitored. If the steelwork is found to be in good working order without any visible defects or signs of deterioration it would be re-greased and re-assembled for continued use.

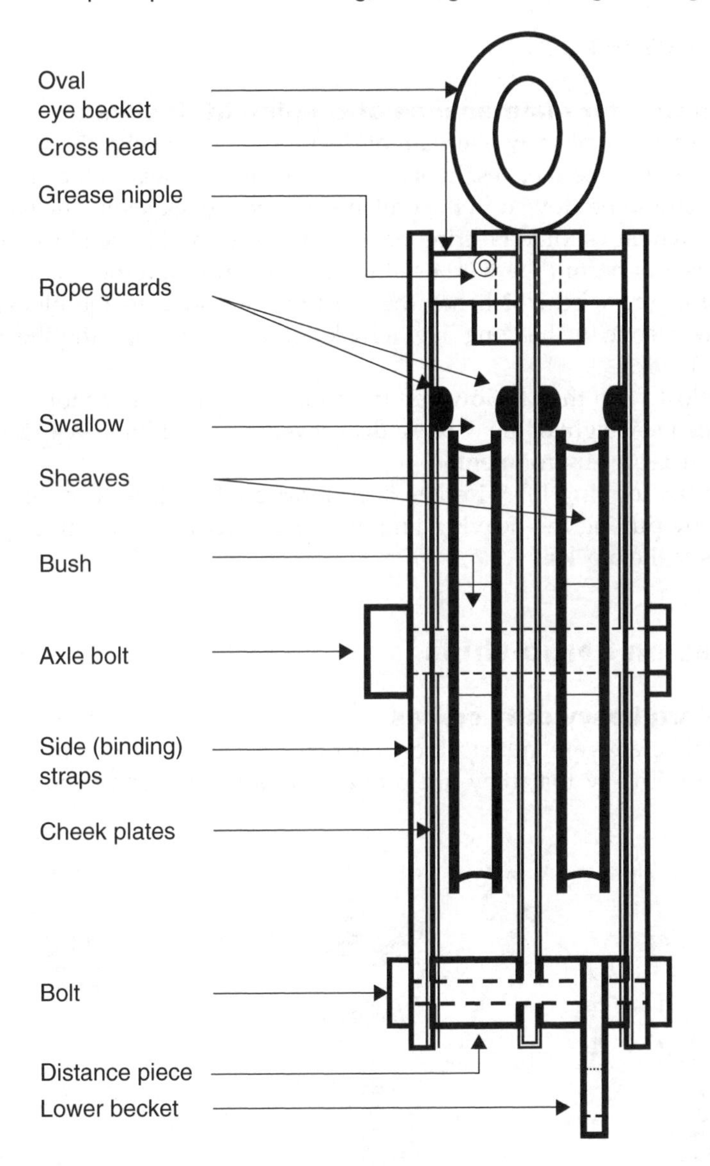

Fig. 1.9 Parts of the Cargo Block.

It is a requirement of the ISM system that lifting gear is correctly maintained and inspected at regular intervals. Most shipping companies comply with this requirement by carrying out such inspections and maintenance under a 'planned maintenance schedule'. Such a procedure ensures that not only lifting gear, but mooring winches, pilot hoists and any other mechanical or weight-bearing equipment is regularly maintained and continuously monitored; inspections, tests and repairs being dated and certificates being retained in the Register of Lifting Appliances and Cargo-Handling Gear.

Deck cranes

Preparation for maintenance of topping lift blocks

Prior to carrying out any overhaul of the topping lift blocks, the wire must be cleared from the sheaves. In order to strip the wire clear of the blocks the derrick should be stowed in the crutch support at deck level. The bare end of the downhaul should be crimped to a cable sock and joined to a heaving line. This will permit the wire itself to be pulled through the sheaves from the end of the wire which has the hard eye shackled to the block. This action will leave the heaving line (long length) rove through the sheaves of the two blocks.

The blocks can then be lowered from the position aloft without bearing the excessive weight of the wire. At deck level the upper blocks can be overhauled in a safe environment.

Once the topping lift wire has been lubricated at deck level, it can be re-rove by pulling the heaving line with the oiled wire back through the sheaves of the blocks.

Cranes on cargo ships

Shipboard heavy duty cranes

To say that cranes are more fashionable than derricks is not strictly a correct statement. To say that they are probably more compact and versatile is

Fig. 1.10 Speed crane/derricks in operation from on top of the Mast House of a general cargo vessel.

more to the point. They tend to be more labour saving than derricks but if comparisons are made for that heavier load capacity and greater lifting capability, then the modern heavy-lift derrick must remain dominant (Figure 1.10).

Single-man drive and control is the key feature of the crane. They can achieve the plumb line quickly and accurately and for up to 40 tonnes SWL they tend to be well suited for shipboard operations. The main drawback for ship-mounted cranes is that the level of shipboard maintenance is increased, usually for the engineering department. They also need skilled labour to handle this increased maintenance workload (Figure 1.11).

In this day and age, flexibility in shipping must be considered essential and such example cranes can be gear shifted into a faster mode of operation

Fig. 1.11 Example of a Deck 25 tonne SWL crane aboard the general cargo vessel Scandia Spirit. The vessel carries two deck cranes, both mounted on the port side of the vessel.

for handling containers up to 36 tonnes or other similar light general cargo parcels (Figure 1.12).

Fig. 1.12 The 'Sir John' general-purpose cargo-vessel lies starboard side to in the Port of Barcelona. Fitted with two heavy-duty deck cranes both situated on the starboard side of the vessel.

In the main, shipboard cranes are in a fixed location, often located offset centre, to one side of the vessel; offset centre cranes having the benefit of an extended outreach for the crane jib. The drawback here is that the vessel is then conditioned to berth crane side to, at each docking, unless working into barges.

Cranes are generally operated with specialized wires having a non-rotational, non-twist property, sometimes referred to as 'wirex'. The lay of the wire being similar to a multi-plate design, wove around a central core in opposition to the directional lay of the core; wires being tested in the normal manner as any other flexible steel wire construction. Despite these anti-twist properties, most incorporate a swivel arrangement over and above the hooking arrangement.

Where cranes are employed in tandem, they tend to be used in conjunction with a bridle or spreader arrangement to engage the total load volume. When lifting close to the crane capacity, such additional items need to be included in the total weight load for the purpose of calculation of the SWL (Figures 1.13 & 2.22).

> ***Note:*** *Some bridle arrangements are often constructed out of steel section and in themselves can add considerable weight to the final load lifted.*

Fig. 1.13 The two deck cranes of the 'Dania' a general cargo/heavy-lift/container option vessel seen lying starboard side to, in Cadiz, Spain. The aft crane is seen in the elevated position while the forward crane is in its stowed position (SWL = 35 tonnes).

Fig. 1.14 The 'Norvik' general cargo vessel pictured in the Port of Limassol, Cyprus. The ship has turned her deck cranes outboard to allow access for the suctions of the grain elevators to discharge a bulk cargo of grain.

Most cranes operate within limits of slew, and with height-luffing limitations. This is not to say that 360° rotational cranes are not available. Virtually all cranes are manufactured to operate through a complete circular arc but limit switches are usually set with shipboard cranes to avoid the jib fouling with associated structures. Safe operational arcs are normally depicted on the ships rigging plan and limit switches are set accordingly (Figure 1.14).

Operator cabs are usually positioned with aerial viewing and provide crane drivers with clear views of the lifting and hoist/ground areas. Topping lift arrangements generally passing overhead and behind the cabin space tend not to interfere with the driver's overall aspects. The hoist and topping lift wires are accommodated on winch barrels found in the base of the crane beneath the cab position.

Crane advantage over derricks	Crane disadvantage over derricks
Simple operation.	Comparatively high installation cost.
Single-man operation, derricks are more labour intensive.	Increased deck space required, especially for 'gantry' type cranes.
Clear deck operational views.	Design is more complex, leaving more to go wrong.
Clear deck space of rigging.	Specialist maintenance required. Hydraulics and electrics.
Versatility with heavy loads, and not required to de-rig.	The SWL of cranes is generally less than that of specialist derrick rigs.
360° slew and working arc when compared with limited operating areas for derricks.	
Able to plumb any point quickly making a faster load/discharge operation.	
Enclosed cabin for operator, where as the majority of derrick operators are exposed, offering greater operator protection and comfort.	
Cranes are acceptably safer to operate because of their simplicity, where derrick rigs can be overly complicated in rigging and operation.	
Cranes can easily service two hatches, or twin hatches in the fore and aft direction because of their 360° slew ability. Derrick rigs are usually designed to service a specific space. *Note: There are exceptions though. Some derrick designs with double-acting floating head rigs can work opposing hatches.*	

Ship's cranes are versatile and have become increasingly popular since their conception. This is because of advanced designs having increased lift capacity and flexible features. They are manufactured in prefabricated steel which incorporate strength section members capable of accepting heavier loads, while at the same time retaining the ability to handle the more regular lighter load.

Gantry cranes (shipboard)

Gantry cranes are extensively found shoreside in the 'container terminals' and these will be described in a later chapter. The use of gantries aboard ships has reduced dramatically on new tonnage because of the extensive facilities found at the terminal ports.

Where gantry rigs do operate, they tend to be 'Tracked Gantry Rigs' which tend to travel the length of the cargo deck in order to service each cargo hold. They also use the rig for moving the hatch covers which are usually 'pontoon covers' that can be lifted and moved to suit the working plan of the vessel when in port.

The gantry structure tends to be a dominant feature and is subject to extensive maintenance attention. However, some small cargo coaster type vessels also use a specific mobile gantry for the sole purpose of lifting off and moving the hatch covers (examples are shown in Figures 1.15 and 1.16).

Fig. 1.15 Example of a mobile tracked gantry crane in operation on the ships foredeck. Suitable for a vessel with all aft accommodation.

Gantry operations

Some shipboard gantry cranes are designed solely to remove and stow pontoon hatch covers while others are suitably employed with outreach capability for working containers to the quayside as well as having the flexibility to remove pontoon covers (Figure 1.17).

Fig. 1.16 A low lying tracked gantry crane operates down the hatch coaming to remove hatch lids from a small coasting vessel carrying bulk cargoes. Single-man operation drives the gantry in a fore and aft direction once the hatch lid is lifted clear of the hatch track.

Fig. 1.17 Dominant gantry crane mounted above the deck and tracked to move fore and aft. Has an SWL of 25 tonnes and outreach extending to 35 m, either side. Single-man overhead operation.

General cargoes – slinging arrangements

Although the majority of cargoes are carried in containers or unitized in one way or another, some cargoes and certainly ship's stores are required to be 'slung' with associated lifting gear. Many bagged cargoes employed 'canvas slings' but handling bagged cargoes proved costly in the modern commercial world and few bagged cargoes are used these days; products being preferred to be shipped in bulk and bagged ashore if required at the distribution stage.

It should be realised that general cargo ships have declined considerably in number, with the main capacity going into the container or Ro-Ro trades. However, some items like pre-slung packaged timber and palletization have gone some way to bridge the ever widening gap between general and containerized cargoes.

Car slings

Single 'private' vehicles are still sometimes loaded and these are crated, containerized or require the customized 'car sling' for open stow. However, where cars (and trucks) are carried in quantity, then 'Pure Car Carriers' (PCCs) or 'Pure Car Truck Carriers' (PCTCs) are normally engaged (Figure 1.18).

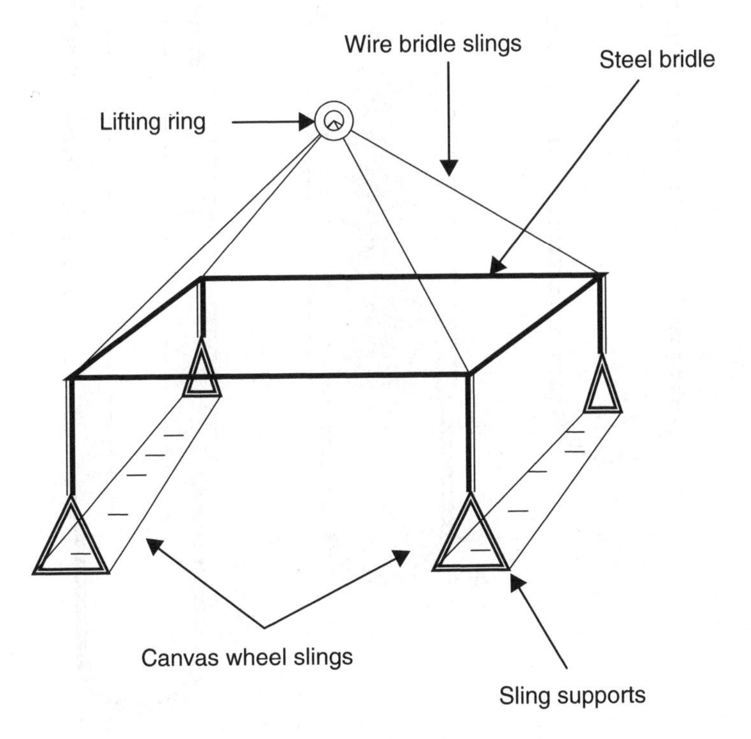

Fig. 1.18 Car sling.

Rope slings

Rope slings are probably the most versatile of slinging arrangements employed in the movement of general cargo parcels. They are made from 10–12 m of 25–30 mm natural fibre rope. Employed for stropping boxes, crates, bales and case goods of varying sizes (Figures 1.19 and 1.20).

The board and canvas slings tend to be specialized for bagged cargo or sacks. With the lack of bagged cargoes being shipped these days, they have

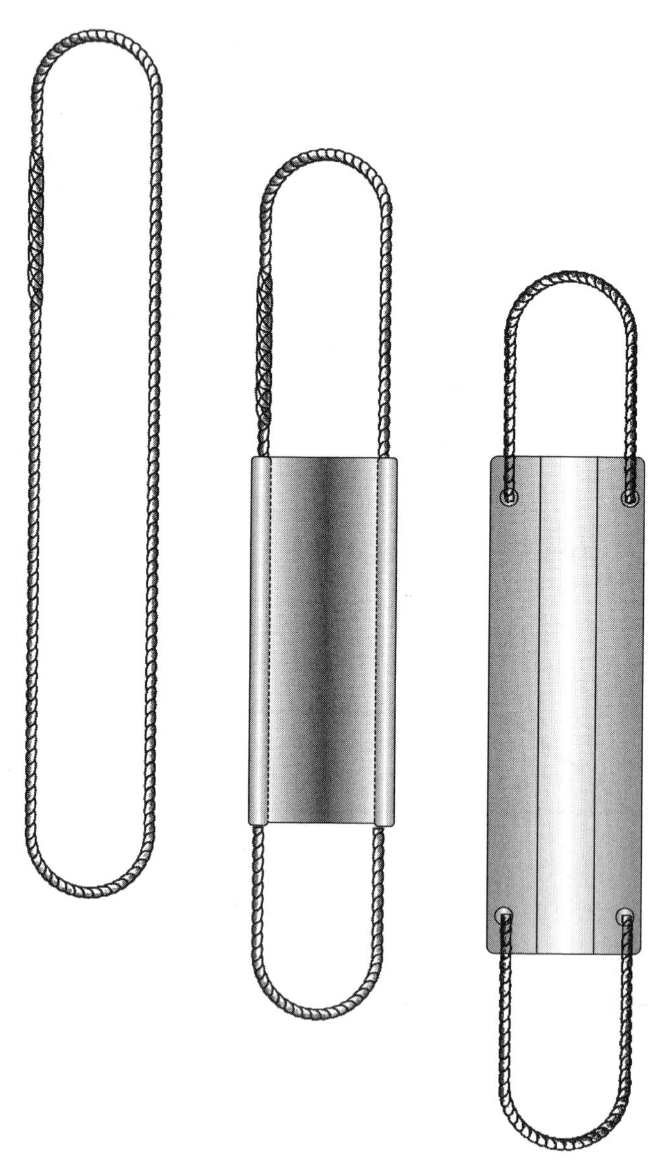

Fig. 1.19 Roped cargo sling arrangements.

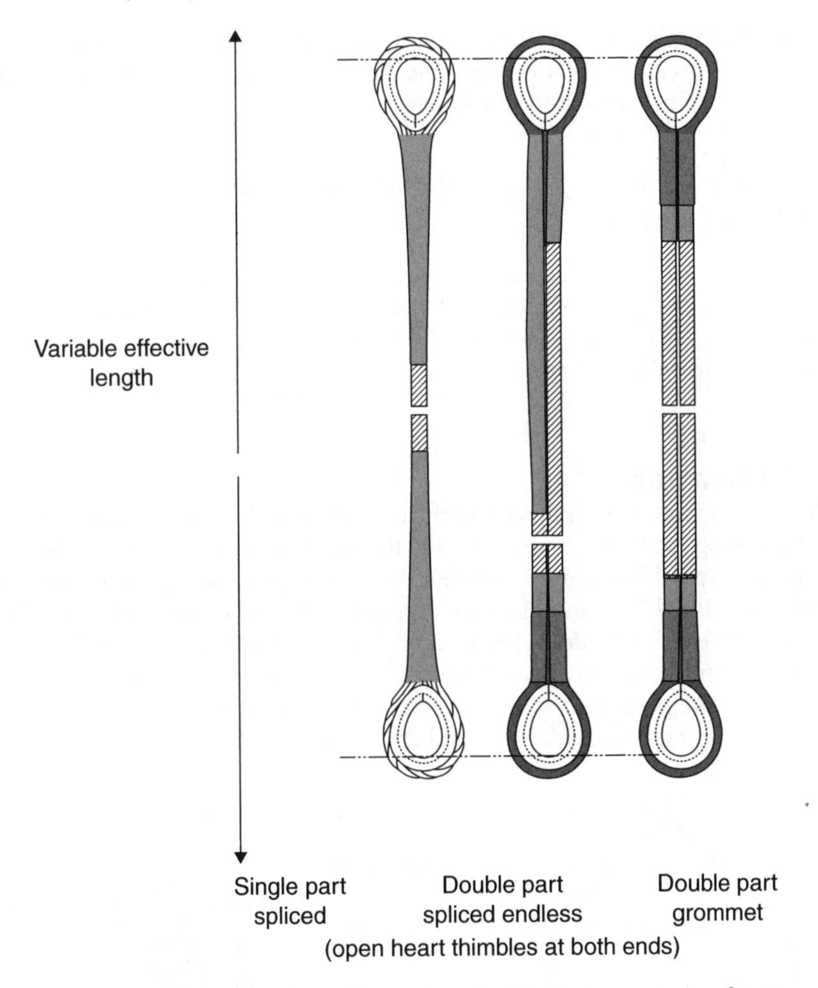

Fig. 1.20 Wire rope slings. Reproduced with kind permission from Bruntons (Musselburgh), Scotland.

dropped away from general use, except in the smaller third world ports; most bagged cargoes now being containerized or shipped on pre-stow pallets.

Multi-legged slings

The permitted working load of a multi-leg sling, for any angle between the sling legs, up to a limit of 90°, is calculated by using the following factors:

2 leg slings	1.25	times the SWL of the single leg
3 leg slings	1.60	
4 leg slings	2.00	

where the angle between the sling legs has limitations and angles of 90° or less are too restrictive, a permissible working load for angle between 90° and 120° can be calculated as follows:

2 leg slings	1.00	times the SWL of the single leg
3 leg slings	1.25	
4 leg slings	1.60	

> ***Note:*** *In the case of three-legged slings the included angle is that angle between any two adjacent legs. In the case of a four-legged sling, the included angle is that angle between any two diagonally opposing legs.*

Palletization

Prior to the massive expansion in the container trade 'palletization' became extremely popular as it speeded up the loading and discharging time of general cargo ships. This meant that the time in port was reduced, together with associated Port and Harbour fees, a fact that was not wasted on shippers and vessel operators. Pre-packed loaded pallets are still widely used around commercial ports and are packed in uniform blocks to minimize broken stowage. Typical cargoes suitable for loading to pallets are cartons, small boxes, crates, sacks and small drums (Figure 1.21).

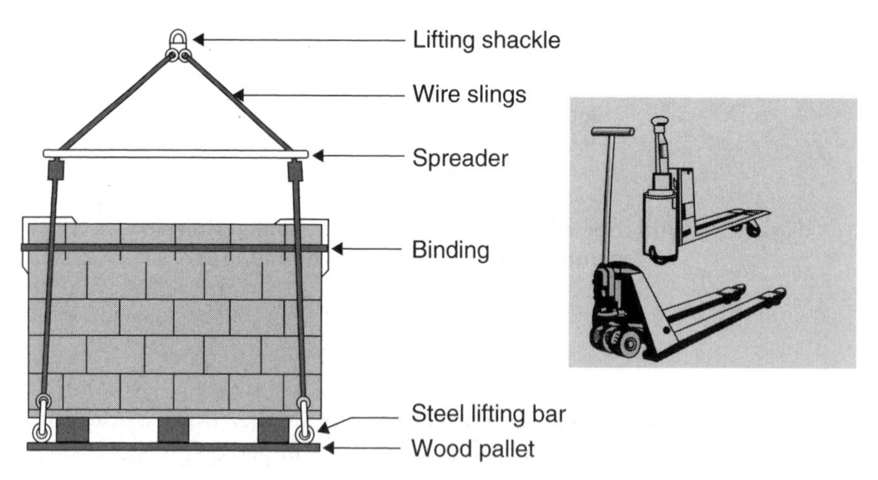

Fig. 1.21 Loaded pallet and pallet transporter.

Palletization has distinct advantages when compared with open stow, general cargo, break bulk-handling methods:

1. less handling of cargo
2. less cargo damage (no hook use and limited pilferage)
3. faster loading discharge times.

Vessels were designed specifically for the purpose of handling pallets and were usually fitted with large open hatchways which allowed spot landing by crane. The ship's design was often multi-deck and fitted with side elevators, shell doors or roll systems to move cargoes into squared-off hatch corners; Tween deck heights being such as to allow access and use of 'fork lift trucks'.

Pallet transporters, battery or manually operated, are useful for 'stuffing' containers where the container floor will generally not have the capacity to support a fork lift truck and its load.

Use of fork lift trucks

The use of pallets and case goods often requires the use of 'fork lift' trucks, either on the quayside or inside the ship's cargo hatches. They have the capability to move cargo parcels out from underdecks into the hatch square to facilitate easy lifting during discharge. Similarly, they can stow heavy individual parcels into a tight stow into hatch corner spaces. It is appreciated that 'bull wires' could be employed for such movements, but rigging and operation of bull wires takes excessive time while the fork lift truck can be effective very quickly. The main disadvantage of fork lift truck use is that the vehicle requires manoeuvring space inside the hatch and, as the hatch is loaded, available space becomes restricted (Figure 1.22).

Fig. 1.22 A fork lift truck operates case goods, stacked timber parcels and palletized drums on the quayside for general cargo vessels.

Ground handling the large load

Fork lift trucks are manufactured in different sizes and are classed by weight. Ship's Officers are advised that the truck itself is a heavy load and will be fitted with a counter weight which provides stability to the working vehicle when transporting loads at its front end. It would be normal practice to separate the counter weight from the truck when lifting it into a ship's hold, especially so if the total combined weight was close to, or exceeded, the SWL of the lifting gear. Once on board the ship the counter weight could then be reunited with the fork lift truck for normal operation (Figure 1.24).

Fig. 1.23 Ground handling of large loads, like containers, can also be achieved by using the larger, high capacity 'fork lift' trucks on the quay side. Expansion forks of extended length are used for wide loads up to about 12 tonnes.

The use of fork lift trucks is a skilled job and requires experienced drivers. Possible problems may be encountered if decks are greasy or wet which could cause loss of traction and subsequent loss of control of the truck when in operation. Spreading sawdust on the deck as an absorbent can usually resolve this situation and keep operations ongoing.

Cargo Officers should exercise caution when working with these trucks aboard the vessel. Although the field of view for the driver is generally good, some cargoes could obscure the total vision and cause blind spots. The nature of the work is such that the number of men inside the hatch should be limited, thereby reducing the possibility of accidents.

Fork lift truck – alternate uses

Probably the most versatile transporters for a variety of cargo parcels that the industry has ever used. The basic fork lift can convert to a mini-crane, drum handler or clamp squeeze tool to suit package requirements. The main forks can be side shifted to work awkward spaces and working capacity can start from 2 tonnes upwards. Height of operations is dependent on the model engaged (Figure 1.24).

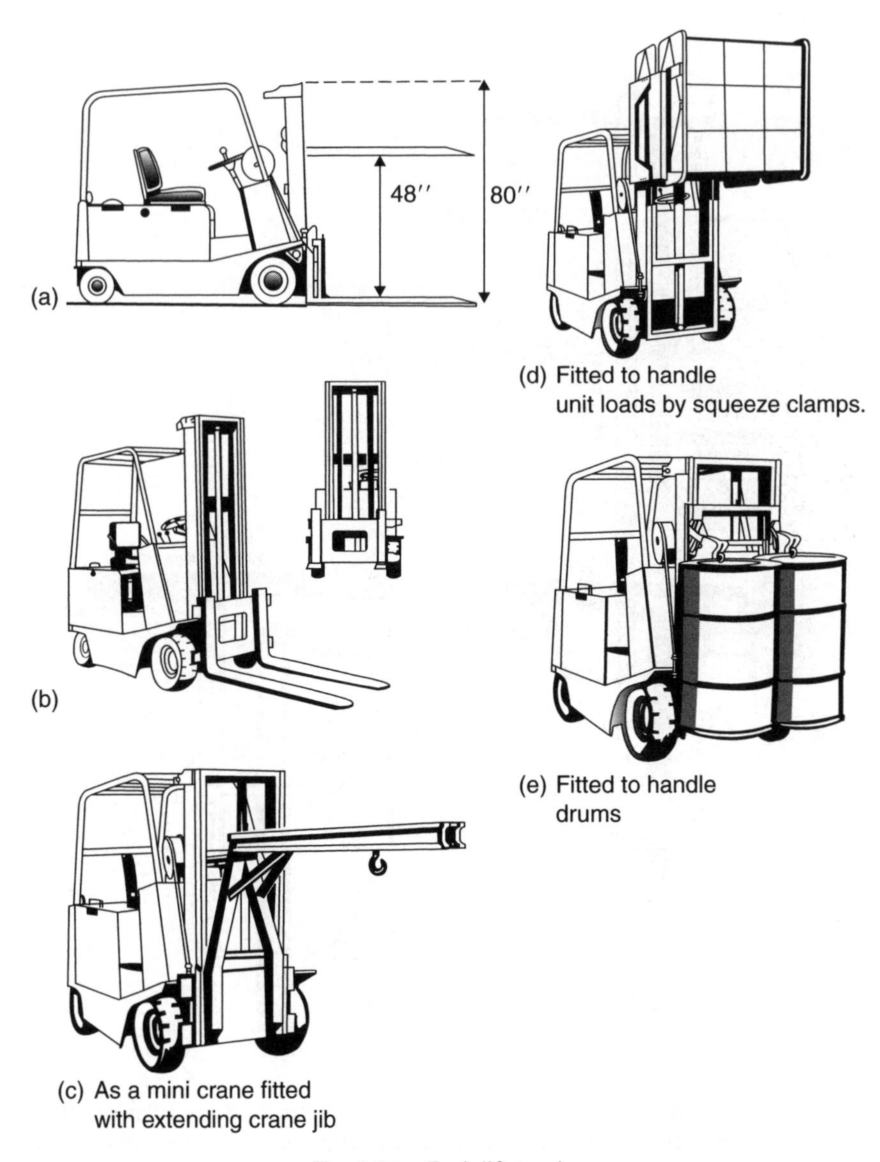

Fig. 1.24 Fork lift trucks.

Side loading practice

Several ships have been constructed with side loading facilities for specific commodities, i.e. paper and forestry products, on the Baltic trades. Watertight hull openings work in conjunction with internal elevators to move cargoes to differing deck levels. These openings, shell doors as such, may function as a loading ramp or platform depending on cargo and designation, fork lift trucks being engaged on board the vessel to position cargo parcels (Figure 1.25).

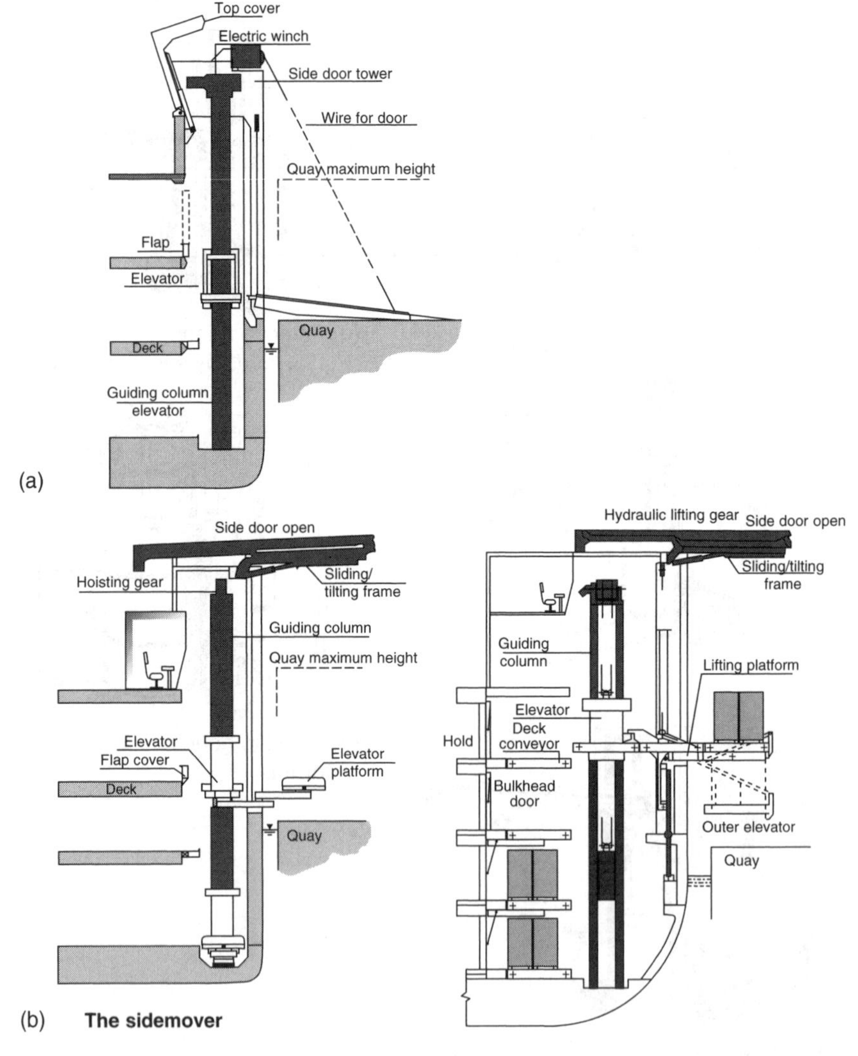

Fig. 1.25 Side loading methods. Reproduced with kind permission from Transmarine, Specialists in Marine Logistics, Equipment, Systems and Services Worldwide.

Chapter 2
Hatchwork and heavy-lift cargoes

Introduction

With the many changing trends of cargo transportation, it would be expected that the design and structure of cargo holds would change to meet the needs of modern shipping. This is clearly evident with container tonnage and the vehicle decks of the Roll-on, Roll-off (Ro-Ro) vessels. However, the changes in the carriage of general cargoes have been comparatively small. This is possible because most merchandise will suit the more popular container or similar unit load movement.

Hold structures have tended to go towards square corners to reduce broken stowage (BS), and suit palletization, pre-slung loads and the use of the fork lift truck inside the holds. Stowage by such vehicles are aided by flush decks in way of the turn of the bilge, as opposed to the angle turn in the sides of the holds of older tonnage.

Some specialist cargoes, like 'steel coils', still suit conventional holds and clearly would not be compatible inside containers, because of the shape and weight of each item. As with large case goods or castings, which tend to transport better by means of conventional stowage in the more conventional type vessel. Such merchandise is clearly edging towards heavy-lift type loads and these heavier loads are covered in detail here, alongside the designated heavy-lift ship and project cargoes.

The objective of this chapter is to provide an overall picture of an industry sector which is an essential part of cargo handling and general shipping practice. It does not have such a high profile as the container or Ro-Ro movement, but it is, nevertheless, an indispensable arm to the practice of shipping.

Hatchwork and rigging (definitions and terminology)

(employed with heavy-lifts and cargo operations)

Backstays – additional strength stays applied to the opposing side of a mast structure when making a heavy lift. These stays are not usually kept permanently rigged and are only set as per the rigging plan when a heavy lift is about to be made.

Bearers – substantial baulks of timber, used to accept the weight of a heavy load on a steel deck. The bearers are laid for two reasons:

1. To spread the load weight over a greater area of deck.
2. To prevent steel loads slipping on the steel deck plate.

Breaking strength – defined by the stress necessary to break a material in tension or compression. The stress factor is usually obtained by testing a sample to destruction.

Bridle – a lifting arrangement that is secured to a heavy load to provide a stable hoist operation when the load is lifted. Bridles may be fitted with a spreader to ensure that the legs of the bridle are kept wide spread so as not to damage the lift and provide a balanced hoist operation.

Bulldog grip (wire rope grips) – screw clamps designed to join two parts of wire together to form a temporary eye or secure a wire end.

Bull wire – (i) a single wire, often used in conjunction with a 'lead block' rigged to move a load sideways off the line of plumb. An example of such a usage is found in dragging cargo loads from the sides of a hold into the hold centre. (ii) a wire used on a single span topping lift, swinging derrick, to hoist or lower the derrick to the desired position. The bull wire being secured to a 'union plate' to work in conjunction with the chain preventor and the down haul of the topping lift span.

Cradle – a lifting base manufactured usually in wood or steel, or a combination of both, employed to accept and support a heavy load. It would normally be employed with heavy lifting slings and shackles to each corner.

Double gear – an expression used when winches are employed in conjunction with making a heavy lift. The purchase and topping lift winches together with any guy winches are locked into 'double gear' to slow the lifting operation down to a manageable safe speed.

Double up – a term used with a derrick which allows a load greater than the safe working load (SWL) of the runner wire but less than the SWL of the derrick, to be lifted safely. It is achieved by means of a longer wire being used in conjunction with a floating block. This effectively provides a double wire support and turns a single whip runner wire, into a 'gun tackle'.

Jumbo Derrick – colloquial term to describe a conventional heavy-lift derrick.

Kilindo rope – a multi-strand rope having non-rotating properties and is a type employed for crane wires.

Lateral drag – the term describes the action of a load on a derrick or crane during the procedure of loading or discharging, where the suspended weight is caused to move in a horizontal direction, as opposed to the expected vertical direction. The action is often prominent when the ship is discharging a load. As the load is passed ashore the ship has been caused to heel over towards the quayside. As the load is landed, the weight comes off the derrick and the ship returns to the upright causing the derrick head to move off the line of plumb. This change of plumb line causes the lifting purchase to 'drag' the weight sideways, e.g. lateral drag.

Lead block – a single sheave block secured in such a position as to change the direction of a weight-bearing wire. Snatch blocks are often used for light working engagement.

Lifting beam – a strength member, usually constructed in steel suspended from the lifting purchase of a heavy-lift derrick when engaged in making a long or wide load lift. Lifting beams may accommodate 'yokes' at each end to facilitate the securing of the wire slings shackled to the load.

Limit switch – a crane feature to prevent the jib outreach from working beyond its operational limitations.

Load density plan – a ships plan which indicates the deck load capacity of cargo space areas of the ship. The Ship's Chief Officer would consult this plan to ensure that the space is not being overloaded by very dense, heavy cargoes.

Maximum angle of heel – a numerical figure usually calculated by a Ship's Chief Officer in order to obtain the maximum angle that a ship would heel when making a heavy lift, to the maximum outreach of the derrick or crane, prior to the load being landed.

Overhauling – (i) an expression used to describe the correct movement of a block and tackle arrangement, as with the lifting purchase of a heavy-lift derrick. The term indicates that all sheaves in the block are rotating freely and the wire parts of the purchase are moving without restriction. (ii) this term can also be used to describe a maintenance activity as when stripping down a cargo block for inspection and re-greasing. The block would be 'overhauled'. (*Note*: the term overhauling is also used to express a speed movement of one ship overtaking another.)

Plumb line – this is specifically a cord with a 'plumb-bob' attached to it. However, it is often used around heavy-lift operations as a term to express 'the line of plumb' where the line of action is the same as the line of weight, namely the 'line of plumb'.

Preventor – a general term to describe a strength, weight bearing wire, found in a 'Union Purchase' Rig on the outboard side of each of the two derricks. Also used to act as support for a mast structure when heavy lifting is engaged. Preventor Backstays generally being rigged to the mast in accord with the ships rigging plan to support work of a conventional 'Jumbo' Derrick.

Proof load – that tonnage value that a derrick or crane is tested to. The value is equal to the SWL of the derrick/crane + an additional percentage weight allowance, e.g. derricks less than 20-tonne SWL proof load is 25% in excess; derricks 20–50-tonne SWL proof load equals +5 tonnes in excess of SWL; derricks over 50-tonne SWL proof load equals 10% in excess of SWL.

Purchase – a term given to blocks and rope (Wire or Fibre) when rove together. Sometimes referred to as a 'block and tackle'. Two multi-sheave blocks are rove with flexible steel wire rope (FSWR) found in common use as the lifting purchase suspended from the spider band of a heavy-lift derrick.

Ramshorn Hook – a heavy duty, double lifting hook, capable of accepting slings on either side. These are extensively in use where heavy-lift operations are ongoing.

Register of ships lifting appliances and cargo handling gear – the ships' certificate and approvals record for all cargo handling and lifting apparatus aboard the vessel.

Saucer – alternative name given to a collar arrangement set above the lifting hook. The function of the saucer is to permit steadying lines to be shackled to it in order to provide stability to the load, during hoisting and slewing operations. They can be fixed or swivel fitted. (*Note*: The term is also employed when carrying 'grain cargoes' where the upper level of the grain cargo is trimmed into a 'saucer' shape.)

Steadying lines – cordage of up to about 24 mm in size, secured in adequate lengths to the load being lifted in order to provide stability and a steadying influence to the load when in transit from quay to ship or ship to barge. Larger, heavier loads may use steadying tackles for the same purpose. However, these are more often secured to a collar/saucer arrangement, above the lifting hook, as opposed to being secured to the load itself. Tackles are rove with FSWR, not fibre cordage.

Stuelcken mast and derrick – trade name for a heavy-lift derrick and supporting mast structure. The patent for the design is held by Blohm & Voss A.G. of Hamburg, Germany. This type of heavy lifting gear was extremely popular during the late 1960s and the 1970s with numerous ships being fitted with one form or other of Stuelcken arrangement.

Tabernacle – a built bearing arrangement situated at deck level to accept the heel of a heavy-lift derrick. The tabernacle allows freedom of movement in azimuth and slewing from Port to Starboard.

Cargo vessel

Modern trend, cargo hold construction

The more modern vessel, probably operating with cranes, may be fitted with twin hatch tops to facilitate ease of operation from both ends of a hold,

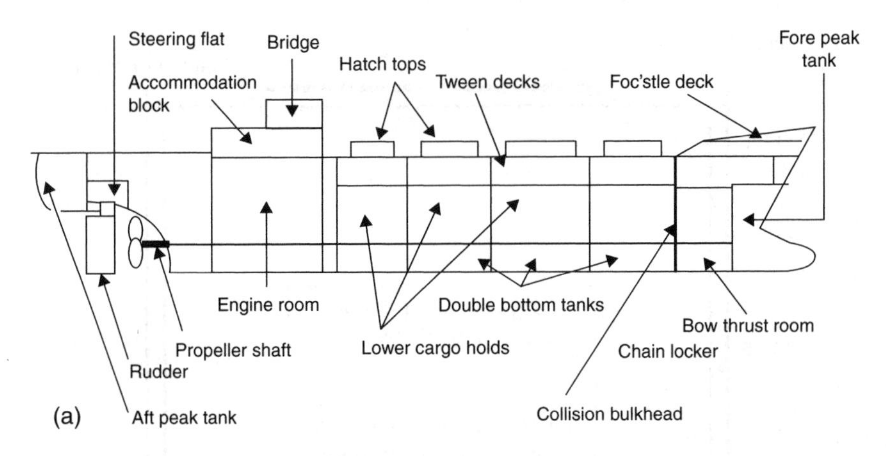

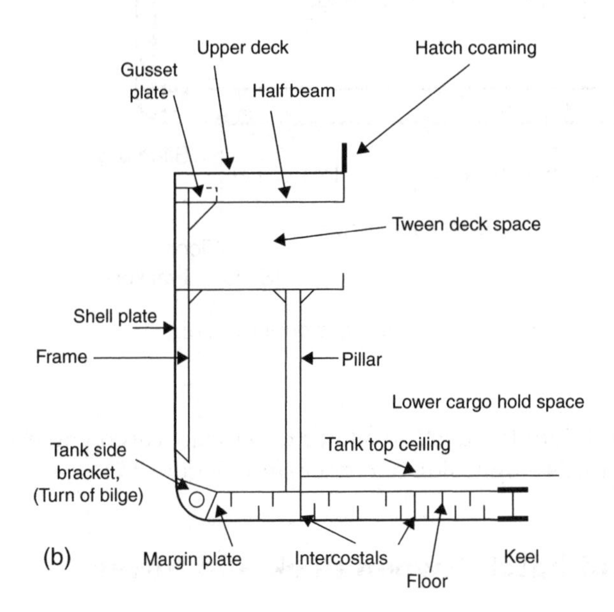

Fig. 2.1 Conventional ship design. (a) general cargo vessel; (b) athwartship – half profile.

while the construction of the hold tends to be spacious to accept a variety of long cargoes. Double hold space with or without temporary athwartships bulkheads which can section the hold depending on the nature of the cargo, provide flexibility to accommodate a variety of cargo types. Figure 2.1(a) shows the conventional ship design of a general cargo vessel. Figure 2.1(b) shows a half profile of the athwartships bulkhead.

Square corner construction lends to reducing BS especially with containers, pallets, vehicles or case goods. Flush 'bilge plate access' is generally a feature of this type of design. Where steel bilge covers (previously limber

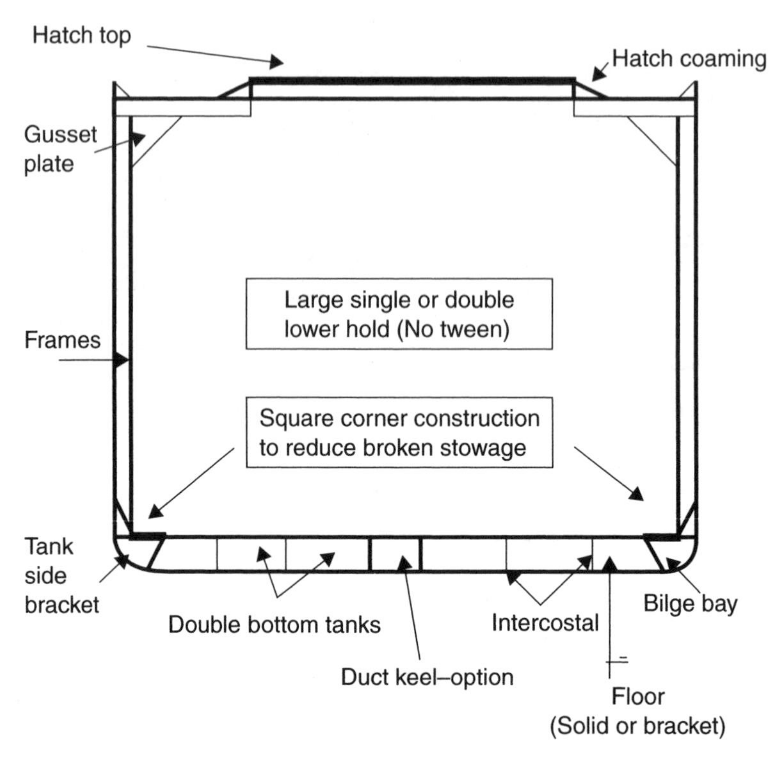

Fig. 2.2 Modern trend, cargo hold construction.

boards) are countersunk into the deck so as not to obstruct cargo parcels being manoeuvred towards a tight side or corner stow (Figure 2.2).

The conventional hatch (tween deck and lower hold (L/H))

An example of the conventional hatch in a general cargo ship is shown in Figure 2.3. This type of hatch was previously covered by wooden hatch boards or slabs but these have been superseded by steel hatch covers. Operated by mechanical means (single pull chain types) or folding 'M types' (hydraulic operation).

Hatch covers

Direct pull (Macgregor) weather deck hatch covers

Figure 2.4 shows a direct pull weather deck hatch cover operation. In this diagram, all hatch top wedges and side locking cleats removed and the tracks are seen to be clear. The bull wire and check wire would be shackled

Fig. 2.3 General cargo seen at the after end of the L/H, while the pontoon tween deck covers are sited stacked in the fore end of the tween deck. Exposed dunnage lies at the bottom of the hold where cargo has been discharged and cargo battens can be seen at the sides of the hold. Safety guard wires and stanchions are rigged around the tween deck in compliance with safety regulations.

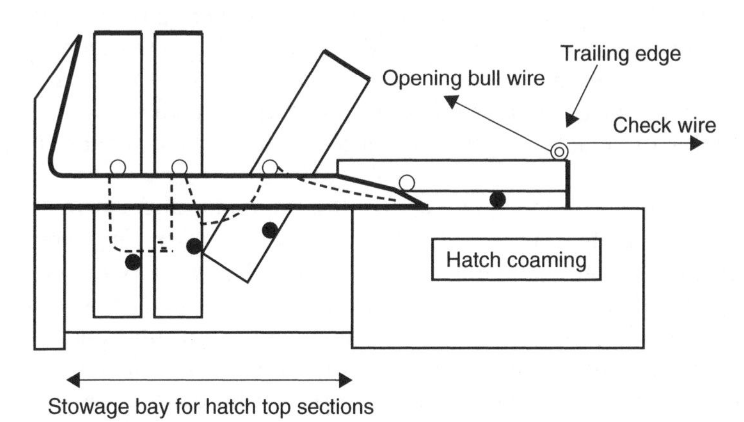

- ● Eccentric wheels lowered to track by manual levers or hydraulics.
- ○ Stowage bay wheels with interconnecting chain.

Raising and lowering of the eccentric wheels by use of portable hand operated jacks or hand levers.

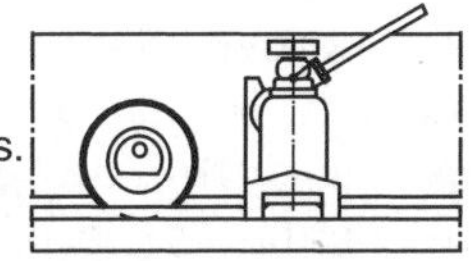

Fig. 2.4 Direct pull weather deck hatch cover. Inset reproduced with kind permission from MacGregor and Co.

to the securing lug of the trailing edge of the hatch top. (*Note*: The bull wire and check wire change function depending on whether opening or closing the hatch cover.) The eccentric wheels are turned down and the 'stowage bay' is sighted to be clear. The locking pins at the end of the hatch would be removed as the weight is taken on the bull wire to open the hatch. Once the hatch lids are open and stowed vertical into the stowage bay, the sections would be locked into the vertical position by lock bars or clamps, to prevent accidental roll back.

Weather deck hatch covers

Steel weather deck hatch covers now dominate virtually all sectors of general, bulk and container shipping. Conventional wooden hatch covers have been eclipsed by the steel designs which are much stronger as well as being easier and quicker to operate. The advantages far outweigh the disadvantages in that continuity of strength of the ship is maintained throughout its overall length. Better watertight integrity is achieved and they are labour saving, in that one man could open five hatches in the time it would take to strip a single conventional wooden hatch. The disadvantages are that they are initially more expensive to install, and carry a requirement for more levels of skilled maintenance.

Once cleated down, a hard rubber seal is created around the hatch top perimeter providing a watertight seal, on virtually all types of covers. Hydraulically operated covers cause a pressure to generate the seal, while mechanical cleating (dogs) provide an additional securing to the cargo space below. The engineering department of the ship usually cater to the maintenance of the hydraulic operations and the draw back is that a hydraulic leak may occur due to say a burst pipe, which could cause subsequent damage to cargo.

Extreme caution should be exercised when opening and closing steel covers, and adequate training should be given to operators who are expected to engage in the opening and closing of what are very heavy steel sections. Check wires and respective safety pins should always be applied if appropriate, when operating direct pull types. Hydraulic folding 'M types' incorporate hydraulic actuators with a non-return capacity which prevents accidental collapse of the hatch tops during opening or closing. Whichever type is employed, they are invariably track mounted and such tracks must be seen to be clear of debris or obstruction prior to operation (Figure 2.5).

Strong flat steel covers lend to heavy lifts and general deck cargo parcels and have proved their capability with the strengthened pontoons which are found in the container vessels. The pontoons having specialized fittings to accept the deck stowage of containers over and above the cargo hold spaces. Similarly, specialized heavy-lift vessels have adopted strengthened open steel decks in order to prosecute their own particular trade sector (Figure 2.6).

Fig. 2.5 Folding hydraulic operated, steel hatch covers, seen in the vertical open position. Securing cleating seen in position prevents accidental roll back.

Fig. 2.6 Rack and pinion horizontal stacking steel hatch covers seen in the hatch open position. The drive chain running the length of the hatch tracks.

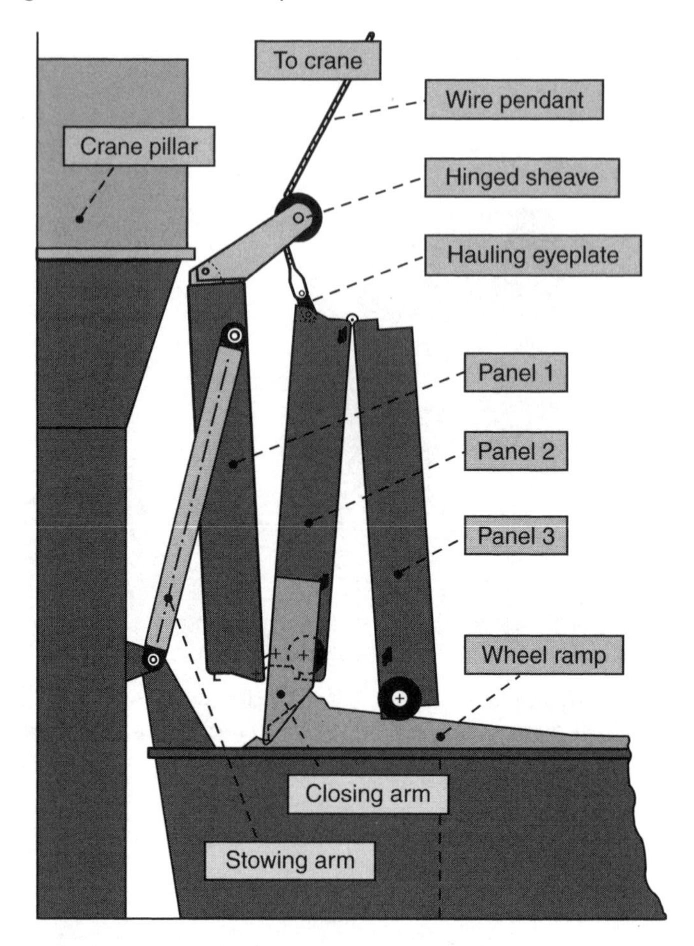

Fig. 2.7 Alternative weather deck hatch cover. Direct pull type. Reproduced with kind permission from MacGregor.

An alternative arrangement is possible when space is not available outboard of the hatches for the deck mounted closing pedestal, e.g. between twin hatches. With this alternative the closing arm operates above coaming level. A wheel ramp is necessary to assist in the initial self-closing action of the covers (Figure 2.7).

Folding (hydraulic operated) hatch covers

The more modern method of operating steel hatch covers is by hydraulics, opening the sections in folding pairs, either single, double or triple pair sections (Figures 2.8 and 2.9).

Multi-folding weather deck hatch covers (MacGregor type)

There are several manufacturers of steel hatch covers and they all generally achieve the same function of sealing the hatchways quickly. Operationally,

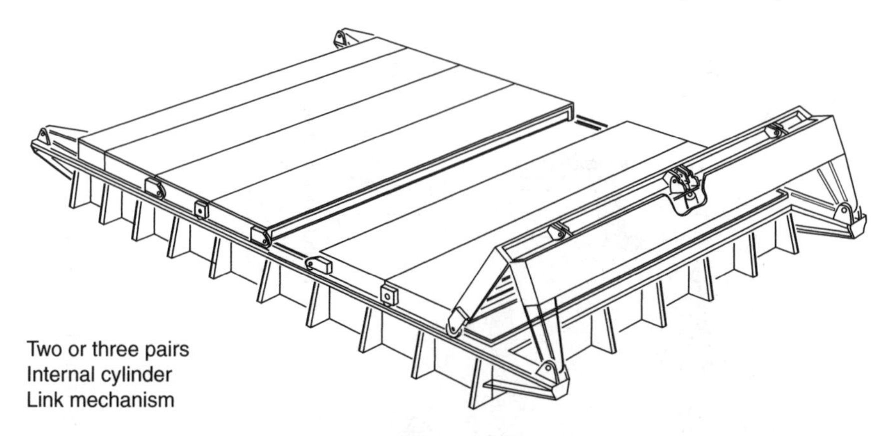

Fig. 2.8 Folding (hydraulic operated) hatch covers. Reproduced with kind permission from MacGregor.

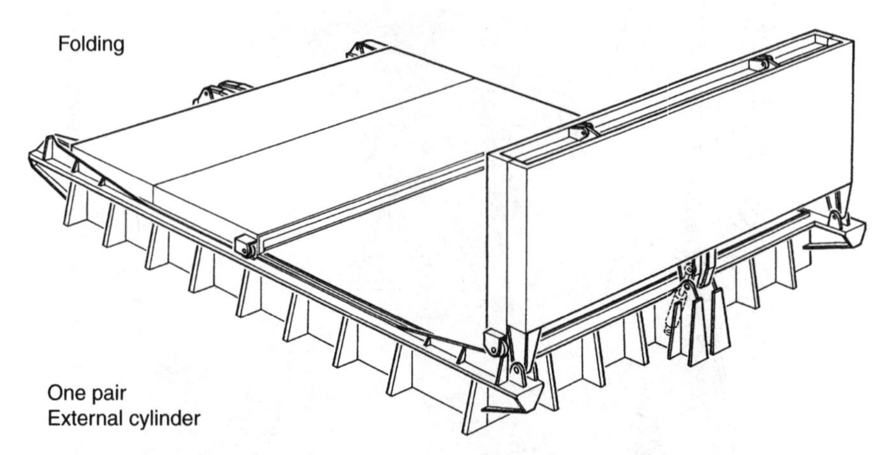

Fig. 2.9 Single pair hatch cover. Hydraulic operated by single external cylinder. Reproduced with kind permission from MacGregor.

one man could close up five or six hatches very quickly by switching on the hydraulic pumps, releasing the locking bars to the stowed sections and operating the control levers designated to each set of covers.

The main disadvantage of hydraulic operations is that the possibility of a burst pipe is always possible, with subsequent cargo damage due to hydraulic oil spillage.

Single pull fixed chain hatch covers

These are automated covers with self drive by built in electric motors (see inset, Figure 2.10). All operations for open and closing the hatch are by push button control. Inclusive of raising the lowering of the covers and operation of the cleating. If desired, these covers can be supplied with sufficient

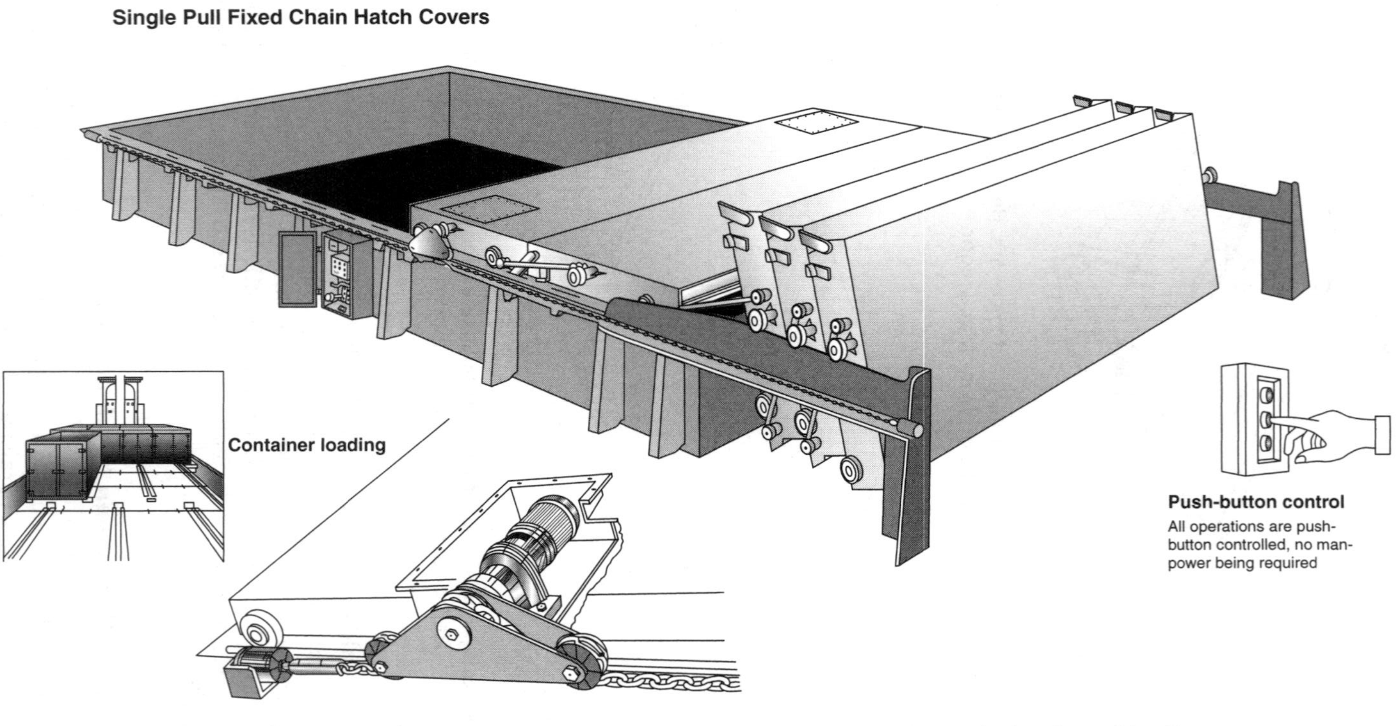

Fig. 2.10 Single pull fixed chain hatch cover. Reproduced with kind permission from MacGregor.

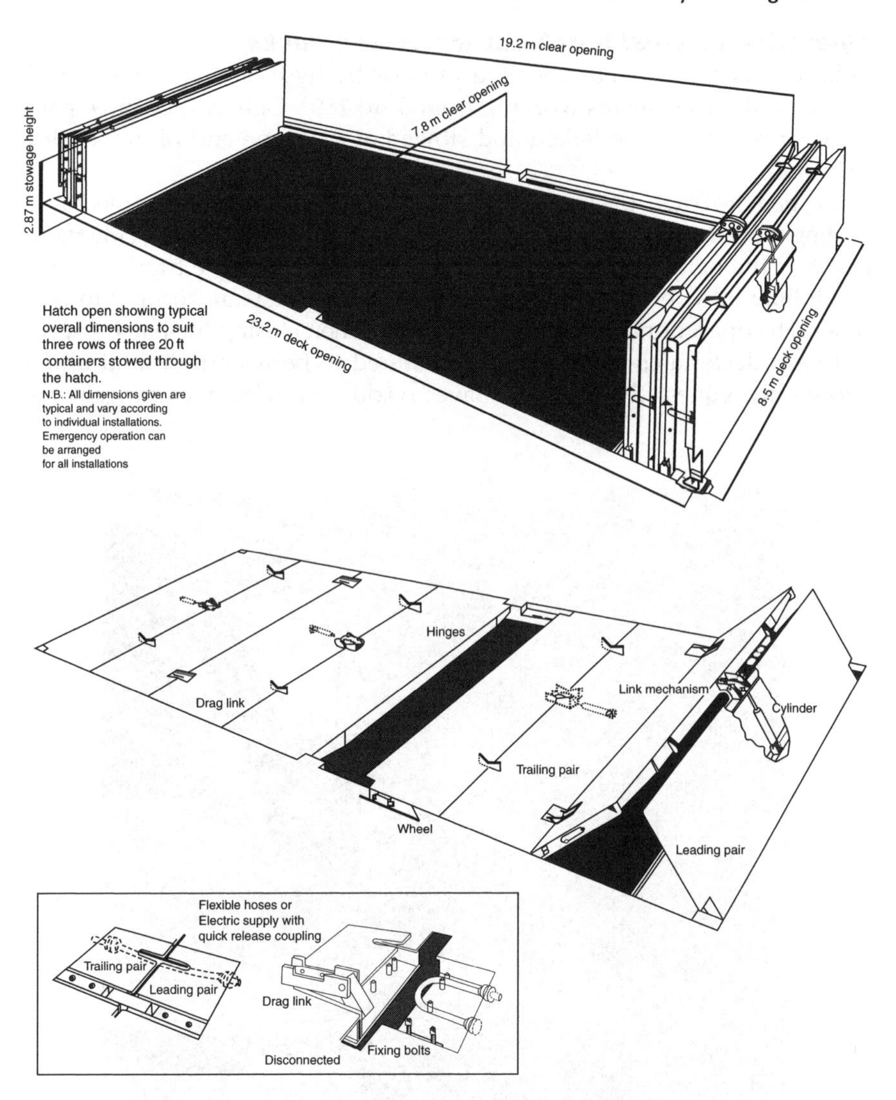

Fig. 2.11 Tween deck 'M type' hydraulic folding hatch covers. Reproduced with kind permission from MacGregor.

strength and the necessary container location sockets to permit the load on top of deck-mounted containers.

Tween deck 'M-type' hydraulic-folding hatch covers

These covers provide a flush and strong deck surface which is ideal for the working of fork lift trucks inside tween deck spaces. Hydraulically operated and user-friendly (Figure 2.11).

Operation of steel hatch covers (tween decks)

Folding hatch covers are operated in pairs by hydraulic cylinders which actuate link mechanisms working from 0° to 180°. One, two or three pairs of cover panels can be linked and stowed at the same end of the hatch if required.

To open the hatch, the leading pair of covers is first operated, immediately pulling the remaining pair(s) into a rolling position on the recessed side tracks and tow the hatch end. Once the leading pair is raised the trailing pair(s) can follow into vertical stowage positions where they are secured to each other. The operational sequence is reversed when closing the hatch covers.

Tween deck hatch covers are not required to be watertight and unless specifically requested they would have no additional cleating arrangements.

Fig. 2.12 Two bulk carrying 'feeder' coasting vessels lie port side to, alongside the grain elevator in Barcelona. The lead ship is seen discharging with partially opened steel hatches, operated by its own mini-gantry crane. The one astern has folding weather deck hatches in the closed-up position.

Partial opening

As with many types of steel covers partial opening is a feature a
achieved comparatively quickly by the operation of quick release 'd
(Figure 2.12).

Loading and discharging heavy lifts

It is normal sea going practice for the Chief Officer of the vessel to supervise the movement of heavy lifts, both in and out of the vessel. This is not, however, to say that the Mate of the ship will not delegate specific functions to the more Junior Cargo Officer or to the stevedore supervisor.

Prior to commencing the lift, the derrick and associated lifting gear needs to be prepared. Many vessels are now fitted with the large 'Stulken-type' derricks, or specialized Hallen or Velle derricks as opposed to the more conventional 'Jumbo' Derrick. Manufacturers' instructions and reference to the ship's rigging plan should always be consulted regarding the preparations of setting up the lifting gear, especially when officers are unfamiliar with the style of rig.

Where a load is outside the SWL of a ship's gear, either a floating crane or a specialized heavy-lift vessel would be employed.

> ***Note:*** *If loading a weight by means of a floating crane, Chief Officers must check that the port of discharge has equivalent lifting apparatus, on the basis that the ship's gear will not be viable for discharge.*

Preparation time for the derrick can vary depending on the type, but a period of up to 2h would not be unusual. Man-management of the rigging crew and advance planning with regard to the number of lifts and in what order they are to be made, in relation to the port of discharge and order of reception of cargo parcels, would be the expected norm.

Stability detail

It must be anticipated that the vessel will go to an angle of heel when making the lift with the derrick extended. This angle of heel should be calculated and the loss of metacentric height ('GM') ascertained prior to commencing the lift. Clearly, any loss of positive stability should be kept to a minimum and to this end any frees surface effects in the ship's tanks should be eliminated or reduced wherever possible.

Operation

Adequate manpower should be available in the form of competent winch drivers and the supervising controller. Winches should be set into double gear for slow operation and steadying lines of appropriate size should be secured to points on the load to allow position adjustments to be made.

Heavy-lift cargoes

When loading or discharging heavy-lifts Deck Officers should be aware of the following precautions and procedures:

1. The stability of the vessel should be adequate and the maximum angle of heel should be acceptable. All free surface effects (FSE) should be eliminated by either 'pressing up' or 'emptying' tanks.
2. If a conventional 'Jumbo' Derrick is employed, then the rigging plan should be referred to with regard to the positioning of 'Preventer Backstays' to support any mast structure.
3. A careful check on the condition of the derrick and associated gear should be made before commencing the lift. Particular attention should be paid to the SWL of shackles, blocks and wires.
4. Ensure all the ship's moorings are taut and that men are standing by to tend as necessary. Fenders should be pre-rigged and the gangway lifted clear of the quayside.
5. All cargo winches affecting the load should be placed in 'double gear'.
6. The deck area where the load is to be landed should be clear of obstructions, and heavy bearers laid to accept and spread the deck weight.
7. The ship's deck capacity plans should be checked to ensure that the deck space is capable of supporting the load.
8. The winch drivers and controller should be seen to be competent, and all non-essential personnel should be clear of the lifting area.
9. Any ship's side rails in the way of the load should be lowered or removed and any barges secured to the ship's side should be cast off.
10. Steadying lines should be secured to the load itself and to the collar of the floating block if fitted.
11. All relevant heads of departments should be advised before commencing the lift.
12. Use the designated lifting points and take the weight slowly. Stop, and inspect all round once the load clears the deck, before allowing the lift to continue.

Examples of slinging arrangements – heavy lifts

Weight and bulk often go together and many of the maritime heavy lifts are not only heavy in their own right but are often extremely bulky by way of having a large volume. Numerous methods have been employed over the years in order to conduct lifting operations in a safe manner. Many types of load beams and bridle arrangements have been seen in practice as successful in spreading the overall weight and bringing added stability to the load movement during a load/discharge activity (Figures 2.13–2.16).

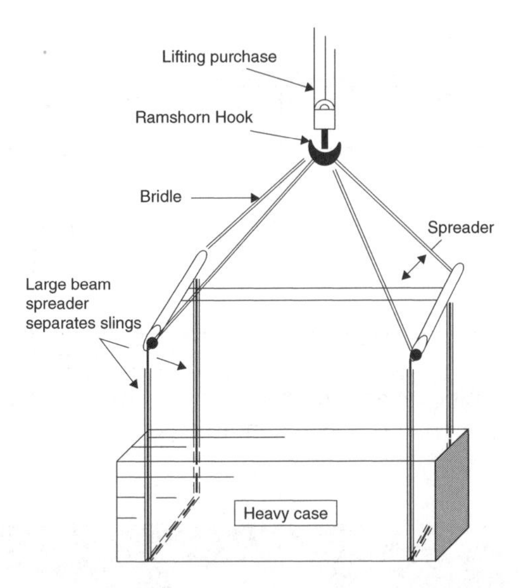

Fig. 2.13 Heavy-lift slinging arrangement.

Fig. 2.14 Use of heavy duty lifting beams. Two Huisman shipboard cranes (each at 275-tonne SWL) hoist the new ferry load Fiorello, by means of two heavy duty lifting beams, aboard the Mammoet vessel 'Transporter'.

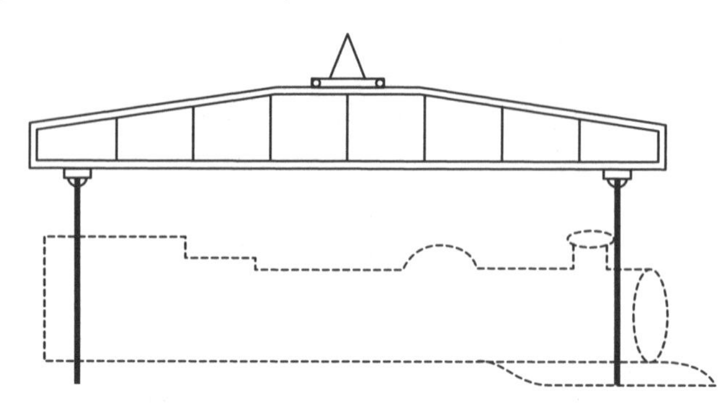

Fig. 2.15 Typical lifting beam employed for long heavy-lift load in the form of a locomotive.

Fig. 2.16 A nuclear waste flask (weighs up to 120 tonnes) is loaded to a customized low load transporter by means of individual heavy duty strops and shackles.

The conventional heavy lift: 'Jumbo' Derrick

In Figure 2.17, a 75-tonne SWL Jumbo Derrick is stowed against the supporting Samson Post structure. The head of the derrick is clamped in the upright position against the upper steel platform. The topping lift is anchored to the underside of the table platform set across the two posts

Fig. 2.17 The conventional heavy lift – a 'Jumbo' Derrick.

(*Note*: The Samson Post structure also supports four conventional 10-tonne SWL conventional heavy Derricks). The terminology and basic working design of a conventional heavy-lift shipboard derrick found up to 150-tonne SWL is shown in Figure 2.18 and for the Jumbo Derrick heavy lift in Figure 2.19.

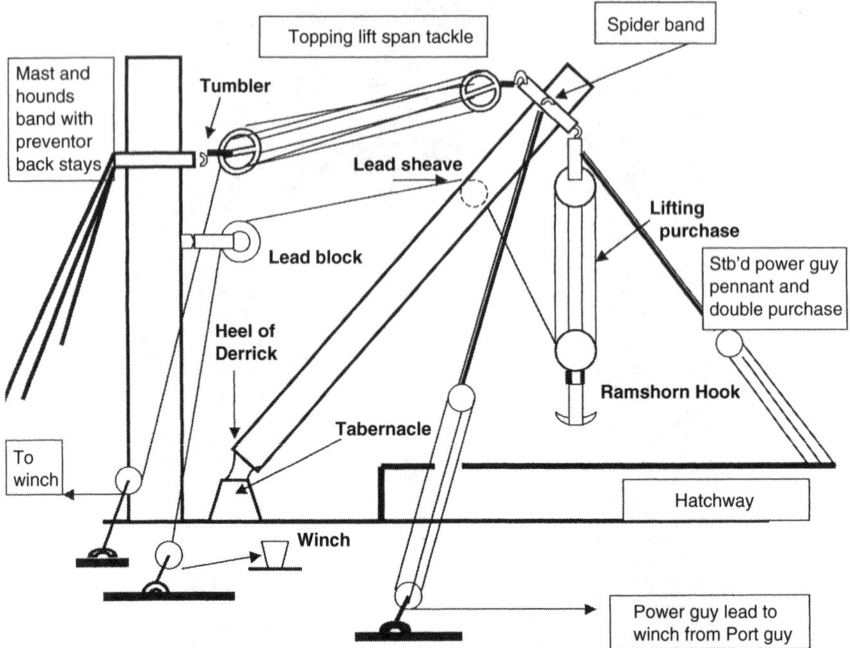

Fig. 2.18 Heavy-lift (conventional) derrick arrangement.

Tandem lifting

It is not unusual these days to encounter specialized vessels, fitted with heavy lift, dual capacity speed cranes. Such ships have the ability to work conventional loads but have the flexibility to load containers or project heavy-lift cargoes Figure 2.20.

Stuelcken derricks

The Stuelcken mast – cargo gear system

The Heavy Lift, Stuelcken systems are noticeable by the prominent angled support mast structure positioned either side of the ship's centre line. The main boom is usually socket mounted and fitted into a tabernacle on the centre line. This positioning allows the derrick to work two hatches forward and aft and does not restrict heavy loads to a single space, as with a conventional derrick.

The Stuelcken Posts, set athwartships, provide not only leads for the topping lifts and guy arrangement but also support smaller 5- and 10-tonne

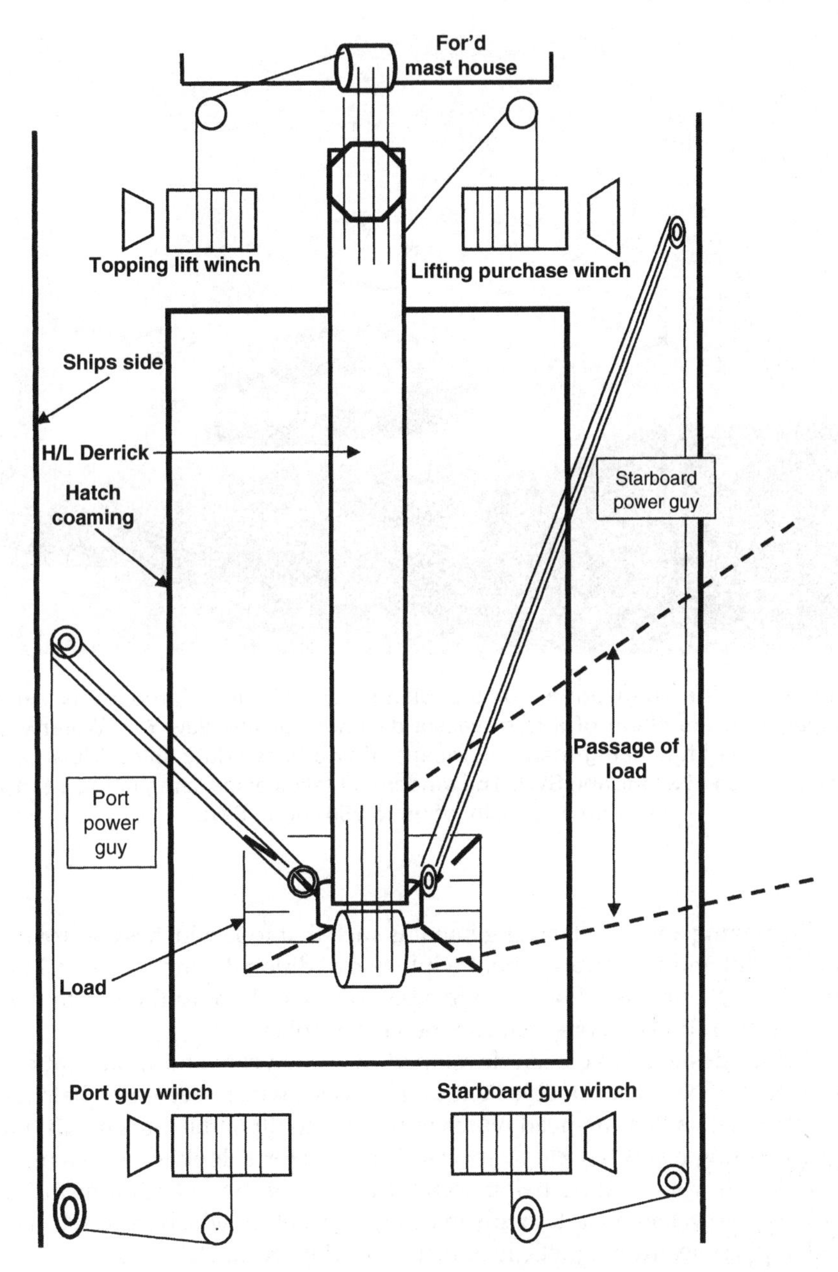

Fig. 2.19 Conventional Heavy-lift 'Jumbo' Derrick arrangement.

derricks with their associated rigging. The posts are of such a wide diameter that they accommodate an internal staircase to provide access to the operator's cab, usually set high up on the post to allow overall vision of the operation.

Fig. 2.20 The 'Transporter' of the Mammoet Shipping Company is seen engaged in the lifting of a ferry vessel destined for the New York Waterway system. The lift is being made by means of two heavy-duty ships 'Huisman' cranes, each of 275-tonne SWL. The tandem lift takes place using lifting beams having a capacity of up to 250-tonne SWL.

The rigging and winch arrangement is such that four winches control the topping lift and guy arrangement while two additional winches control the main lifting purchase. Endless wires pay out/wind on, to the winch barrels, by operation of a one-man, six-notch controller.

Various designs have been developed over the years and modifications have been added. The 'Double Pendulum' model, which serves two hatches, operates with a floating head which is allowed to tilt in the fore and aft line when serving respective cargo spaces. A 'Rams Horn Hook' with a changeable double collar fitting being secured across the two pendulum lifting tackles. The system operates with an emergency cut-off which stops winches and applies electro-magnetic locking brakes (Figure 2.21).

Stuelcken derrick rigs are constructed with numerous anti-friction bearings which produce only about 2% friction throughout a lifting operation. These bearings are extremely durable and do not require maintenance for about 4 years, making them an attractive option to operators.

The standard wires for the rig are 40 mm and the barrels of winches are usually spiral grooved to safeguard their condition and endurance. The length of the span tackles are variable and will be dependent on the length of the boom (Figures 2.22 and 2.23).

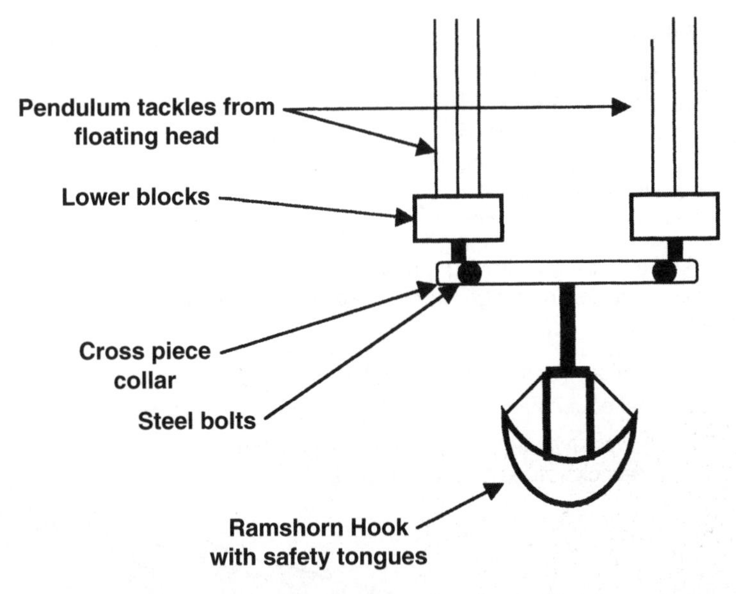

Fig. 2.21 Cargo gear system – double pendulum model.

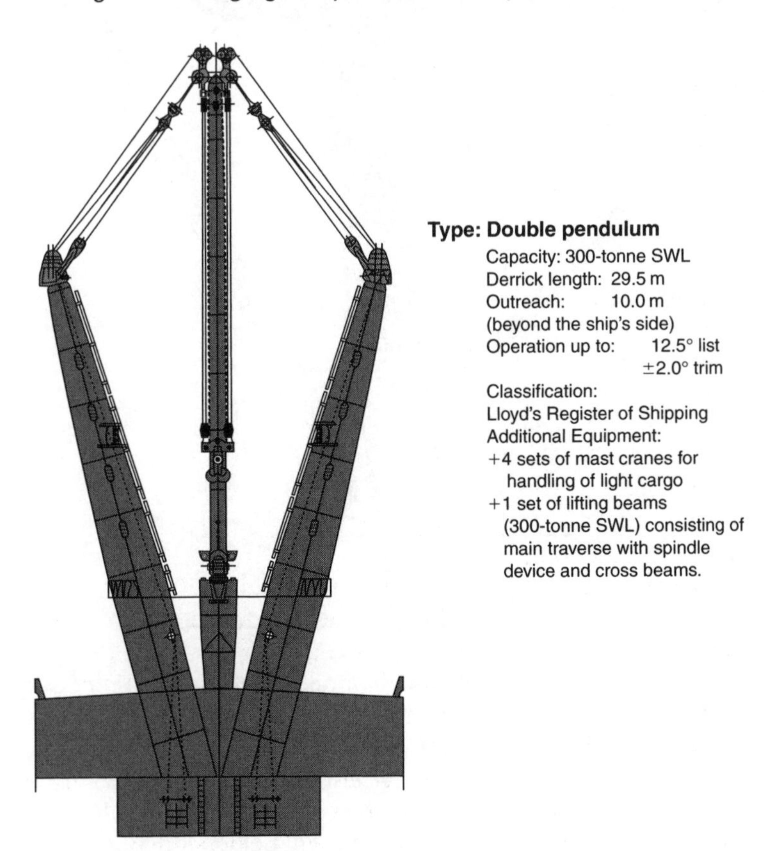

Fig. 2.22 Double pendulum model.

Type: Pivot type

Capacity: 250-tonne SWL
Derrick length: 30 m

Outreach:	beyond portside	14.15 m
	beyond starboardside	10.0 m

Operation up to: 5° list
±2° trim

Designed for three slewing ranges of each 100°. Change of slewing range without load. Total slewing range 260°.

Classification: Germanischer Lloyd

Additional equipment:

+ Each Pivot type equipped with two log/lumber type derricks of each 35-tonne SWL.

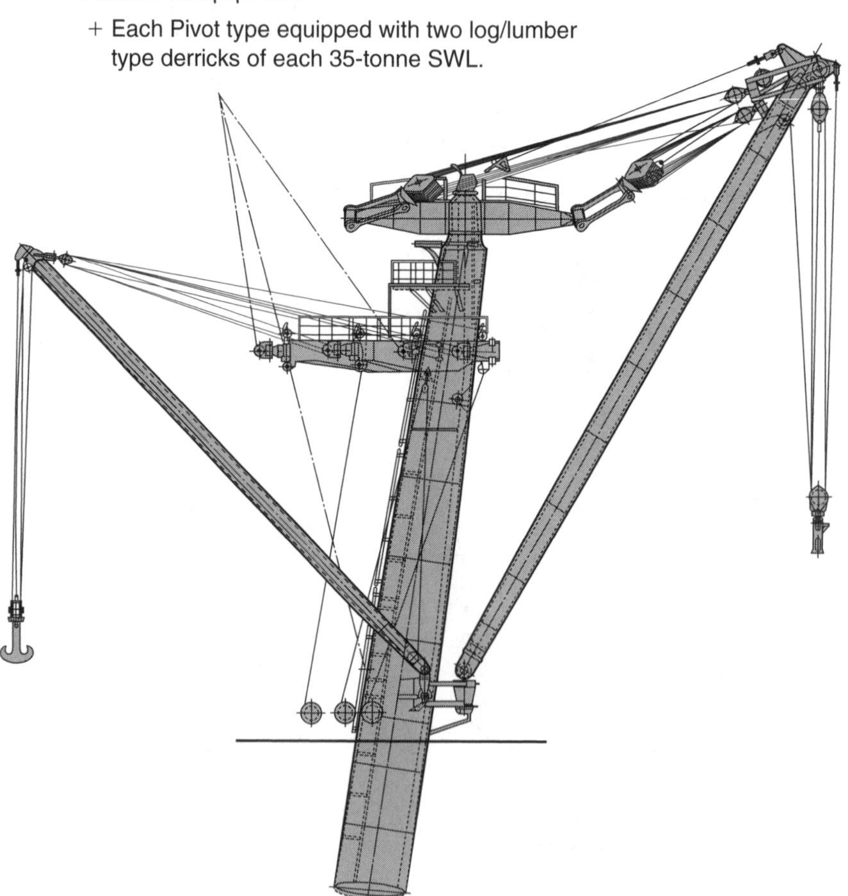

Fig. 2.23 Stuelcken masts and cargo lifting arrangement.

Although Stuelcken rigs still remain operational, their use has diminished with the improved designs of heavy-lift vessels, which previously tended to dominate the 'Project' cargo section of the industry.

Heavy-lift floating crane

Conventional heavy cargo loads, which are scheduled for carriage by sea, are often required to be loaded by means of a floating crane. When the load is too great to be handled by the ship's own lifting gear, the floating crane option is usually the next immediate choice. Most major ports around the world have this facility as an alternative option for heavy specialist work. The type of activity is two-fold, because, if loaded by this means at the port of departure, the same load must be discharged at its destination by similar or equivalent methods. (Ship's Chief Officers need to ascertain that if the load is above the ships lifting gear capability, that the discharge port has means of lifting the load out.)

A Ship's Cargo Officer needs to ensure that the heavy load is accessible and that the floating crane facility is booked in advance in order to make the scheduled lift. Booking of a special crane would normally be carried out via the ship's agents, leaving a ship's personnel very much in the hands of external parties; the Port Authority often controls the movement of all commercial and specialist traffic in and around the harbour.

The 'floating crane' should not be confused with the specialist 'crane barge'. Floating cranes differ in that they may not be self-propelled and may require the assistance of tugs to manoeuvre alongside the ocean transport, prior to engaging in the lift(s). The construction of these conventional cranes is such that the crane is mounted on a pontoon barge with open deck space to accommodate the cargo parcel, the pontoon barge being a tank system that can be trimmed to suit the necessity of the operation if the case requires.

The main disadvantage against the more modern, floating sheer legs, is that generally speaking the outreach of the crane's jib is limited in its arc of operation. Also, the lift capacity can be restrictive on weight when compared with the heavier and larger units which tend to operate extensively in the offshore/shipyard arenas.

When booking the facility, agents need to be made aware of the weight of the load and its overall size; also its respective position on board the vessel, together with its accessibility. Hire costs of the unit are usually quite high and with this in mind, any delays incurred by the ship not being ready to discharge or accept a scheduled load on arrival of the crane, could become a costly exercise.

The crane/sheer leg barge (self-propelled)

Derrick/crane barges tend to work extensively in the offshore sector of the marine industry but their mobility under own propulsion, together with thruster operations, provide flexibility to many heavy-lift options. Some

Fig. 2.24 The Smit 'Cyclone' floating sheer leg barge, engages in a general cargo heavy-lift operation on the vessels offshore side.

builds incorporate dynamic positioning and depending on overall size, have lifting capacity up to and including 6000 tonnes with main crane jib operations (Figure 2.24).

Heavy-lift ships and project cargoes

The need for heavy-lift ships developed alongside the immense size of the loads required within the development of the offshore industry. Its origins probably come from the idea of the 'floating dry dock' which has been around for many years before the offshore expansion. The principle difference between the floating dock and the heavy-lift ship is that one is always self-propelled and acts as a regular means of transportation, while the floating dock is usually annexed to a shipyard and if it is required to move position, such a move would normally be handled by tugs.

They both have operational tank systems which allow them to work employing the same Archimedes principle of flotation. Submerging themselves to allow a load to float in, or over, prior to de-ballasting and lifting the load clear of the water line. The Heavy-Lift Ship generally does not submerge its loading deck more than to a calculated depth, but enough to allow 'float over' methods to operate (Figures 2.25–2.28).

Float over loading methods must therefore be capable of accommodating the draught drawn by the load when waterborne, the actual load usually

Fig. 2.25 The steel deck of the 'Baltic Eider' fitted to receive containers overall. Seen in a part-loaded condition passing through light ice in the Baltic Sea.

Fig. 2.26 The steel deck space of the 'Sea Teal' a heavy-lift transport. The tank ballast system, allows the deck to be submerged to allow a Float-On, Float-Off, system to take place for heavy-lift or project-cargoes.

Fig. 2.27 The heavy-lift vessel 'Super Servant 3' seen loaded with the crane barge AL-Baraka 1, lies at anchor awaiting to discharge its load by Float-off methods.

being rafted and towed or pushed by tugs to a position over the transports deck. Once in position over the load deck, the de-ballast operation of the heavy-lift vessel can take place allowing the deck to rise and so raising the load clear of the surface. The load, complete with raft (if employed), is then transported under the vessel's own power.

Elements for consideration for heavy-lift transports:

1. Overall size-dimensions of the load
2. Weight of the load
3. Weight of lifting accessories
4. SWL of Lifting elements
5. Weather conditions
6. Positive stability of transporting vessel
7. Density of water in load and discharge ports
8. Ballast arrangements for trim and list of vessel
9. Passage plan of transport route
10. Fuel burn on route
11. Speed and ETA of passage
12. Loadline zone requirements not infringed
13. Method of discharge
14. Facilities of discharge Port

Fig. 2.28 Stern view of the heavy-lift vessel 'Super Servant 3' with the crane barge load seen in the Arabian Gulf area.

15. Manpower requirements for loading/shipping/and discharging
16. Documentation for the load
17. Specialist handling personnel
18. Communication facilities to accommodate loading/discharge
19. Securing arrangements for load on route
20. Load management on voyage.

Large heavy loads tend to accrue logistical problems from the time of construction to that moment in time when the load arrives at its final destination. The shipping element of the load's journey is just one stage during the transportation. Cargo surveyors, safety experts, company officials and troubleshooters of various kinds tend to move alongside the passage of the load up to that time of final delivery.

Planning for project cargo transport

It would be natural for the layman to assume that the heavy load just moves on its own with the help of a police escort, but this is clearly not the case for the extreme load, or that larger-than-large plant (Figure 2.29). Planning of the delivery must be known prior to the load being built. A company may be able to build for the customer but if the load cannot be transported safely, because of weight or size, then the actual building becomes a 'white elephant', in more ways than one.

Also costs for the transportation could be considerable and these would expect to influence financial agreements and be included at the contract stage.

Fig. 2.29 Project type cargoes. The Smit 'Giant 4' heavy-lift transport engages with a 6500-tonne lift, designated for the 'Visund Field' offshore. Giant 4, is a 24 000 dwt submersible heavy duty transport barge. The operation was conducted by a Load-out/Float-off arrangement in June 1997.

Transportation – planning considerations, project cargoes

Measurement of the load

Not only weight measurement of the cargo, but its overall length, breadth and depth will be required. If the load is structured to float during the loading or discharging period, then the draught at which flotation occurs together with the freeboard measurement; the centre of gravity (C of G) of the load mass; and if applicable, the centre of buoyancy; density of water at the loading point; and density of the water at the point of discharge, must be calculated. Tidal considerations at the load and discharge positions should also be calculated for the designated periods.

Transport vehicle-considerations

Capability of the carrier to carry out the task......

In the case of a ship, is the vessel capable of accepting the load? What is the displacement and physical size of the vessel and its deck load density capability? What is the metacentric height ('GM'), and what will be the new 'GM' with the load added? Further consideration must be given for the general assessment of the ship's stability throughout all stages of the passage; endurance of the vessel; and the effect of burning bunker oil and consuming water; ballast movement and the ability to trim or list the vessel for the purpose of loading/discharging; number of crew; experience of the master; Charter rates; and not least the availability of the vessel.

Shoreside administration for heavy-lift operations

Every heavy-lift operation will pass through various degrees of administration prior to the practical lift taking place. The manufacturers/shippers will be required to provide clear information as to dimensions, weight, lifting and securing points, and the position of the C of G before the load can be accepted by the ship; while the ship may be required to give details of its crane capability, inclusive of outreach and load capacity.

The loading operation itself as to whether it will be from the quayside, or from a barge, must also be discussed, together with the detail of use of ship's gear or floating crane. Weather conditions and mooring arrangements may also be featured at this time. Once loading is proposed, the stability data and the maximum angle of heel that will be attained would need to be calculated. Ballast arrangements pertinent to the operation may well need to be adjusted prior to contemplating the actual lift.

The ship would no doubt be consulted on voyage and carriage details, as to the securing of the load, the deck capacity to accommodate the load, and the stability criteria. The ship would also require assurances regarding the port of discharge and the capabilities of said port. If the load is beyond the capacity of a ship's lifting gear, then the discharge port must have accessibility to a floating crane facility and that this facility will be available at the required time.

Where road transport is involved in delivering the load to the quayside, road width and load capability would need to be assessed. A 500-tonne load on the back of a low-loader may well cause landslip or subsidence of a roadside, which must be clear of obstructions like bridges and rail crossings. Wide loads or special bulky loads may require police escort for movement on public highways to and from loading/discharge ports.

Once loaded, the weight will need to be secured and to this end a rigging gang is often employed. However, prudent overseeing by Ship's Officers is expected on this particular exercise. Bearing in mind that the rigging gang are not sailing with the ship, and once the ship lets her moorings go, any movement of the load will be down to the ship's crew, to effect re-securing.

Customs clearance would also be required as per any other cargo parcel and this would be obtained through the usual channels when the manifest is presented, to clear the vessel inwards. Export licences, being the responsibility of the shipper, together with any special details where the cargo is of a hazardous nature, covered by special clearances, e.g. armaments.

Movement logistics of the large load

Clearly, the task of transporting 'project' cargoes does not lend itself easily to the use of public roads. Fortunately, the building sites for such items are often located by coastlines and generally do not encroach on public highways. For example, shipyards build and transport modules or installations within their own perimeters and transport within those same perimeters. However, occasionally, that one-off project requires a specialized route. Timing is critical at all stages of the journey to ensure minimal disruption to the general public, and police escort must be anticipated door to door (Figures 2.30 and 2.31).

For the transport of heavy loads (ground handling equipment), further reference should be made to the IMO publication on '*The Safe Transport of Dangerous Cargoes and Related Activities in Port Areas*'.

Voyage planning

The movement of project cargoes is, by the very nature of the task, generally carried out at a slow speed. This is especially so as in the examples shown as extreme lifts on pages 62/65. Tug assistance is often employed and the operation must be conducted at a safe speed for the circumstances, the movement between the loading port of departure, towards the discharge position, being carried out under correct navigation signals appropriate to each phase of the passage.

As with any passage/voyage plan, the principles of 'passage planning' would need to be observed, but clearly specialist conditions apply over and above those imposed on a conventional ship at sea. Passage planning involves the following phases:

Fig. 2.30 Mobile transport platform employed for ground handling of steel installation. This section of the movement plan must be considered prior to the build stage.

Appraisal – the gathering of relevant charts, publications, informations and relevant datas to enable the construction of a charted voyage plan.

Planning – the actual construction of the plan to highlight the proposed route. To provide details of way points, bunkering stations, navigation hazards, margins of safety, currents and tidal informations, monitoring points, contingency plans, traffic focal points, pilotage arrangements, underkeel clearances, etc.

Fig. 2.31 Multi-wheel heavy load transports. Manufactured by the Scheuerle Company of Germany. The mobile platforms provide multi-axle transport for the large heavy load and are regularly inter-connected for the project load carrying up to 10 000 tonnes. Variations for smaller loads range from the 15-tonne low-bed trailers to platform trailers of up to 1000 tonnes.

Execution – the movement of the transport to follow the plan through to its completion. The positive execution of the plan by the vessel.

Monitoring – the confirmation that the vessel is proceeding as per the designated plan. Position monitoring is taking place and the movement of the vessel is proceeding through the various stages of the voyage.

> *Note: A passage plan is equally meant to highlight the areas where the vessel should not go, a particular important aspect to vessels engaged with 'project cargoes'. The load may restrict passage through canals, under bridges or through areas of reduced underkeel clearance (UKC).*

Voyage plan acceptance

Once the plan is constructed, it would warrant close inspection by the Project Manager and the Ship's Master. Such a plan would need to incorporate a considerable number of special features prior to being considered acceptable to relevant parties. Passage plans are made up to ensure 'berth to berth' movement is achieved safely and a plan for movement of a project cargo would expect to include the following special features:

Risk assessment Completed on the basis of the initial plan (passage plans are flexible and circumstances may make a deviation from the proposal to take a necessary action when on route).

Communications Methods: VHF channels; Secondary methods: Advisory contacts, Coastguard, VTS, Hydrographic Office, Meteorological Office, Agents, Medical contingency. Most towing operations and project movements would normally be accompanied by a navigation warning to advise shipping likely to be affected. Such warnings could be effected by Coast Radio Stations, Port and Harbour controls, and/or the Hydrographic Office of the countries involved.

Loading procedures Methods: Various examples: Lift-On/Lift-Off, Float-On/Float-Off, etc. Tug assistance, marine pilots, rigging and lifting personnel as required. Tidal conditions, weather conditions monitored.

Securing procedures Personnel and associated equipment, Surveyor/Project Manager inspection. Contractors: riggers, lashings, welders.
Risk assessment – Tolerable.

Safety assessment LSA/manpower, Navigation equipment test. Engine test. Weather forecast 48 h, long-range forecast.

Route planning Weather, ports of call, mooring facilities, UKC, width of channel. Position Monitoring methods: communications to shore to include progress reports, Navigation hazards, Command Authority, canal passage or bridge obstructions. Traffic focal points. Seasonal weather considerations.

Contingencies Endurance, bunkers, manpower, emergency communication contacts. Weather, mechanical failure, steering failure, tug assistance. Use of anchors, safe anchorages. Special signals. Support services (shore based).

Schedule Timing to effect move, speed of move relevant to each movement phase. Charter Party, delivery date, 'penalty clauses'. Sailing plan, monitoring and tracking operations, progress reports.
Risk assessment – Per phase of voyage.

Discharge procedure Methods: Ground handling equipment, secondary transport. Specialist personnel and equipment. Quayside facilities and tidal considerations. Risk assessment.

Personnel requirements Surveyors, specialist handlers, various contractors.

Insurance – Shoreside administration.

Documentation/Customs clearances Reception, delivery communications, Export licences.

Ancillary units Tugs, Lifting units, equipment, consumables.

Specialist equipment Ice regions.

Accommodations Airports, hotels, local transport facilities, labour force.

Security Piracy, road transport, in port, at sea, communications. Police, military, security codes affecting contingencies.

Costs Market assessment, Political considerations.

Chapter 3
Stowage properties of general cargoes

Introduction

General cargo is a term which covers a great variety of goods. Those goods may be in bags, cases, crates, drums or barrels, or they may be kept together in bales. They could be individual parcels, castings or machinery parts, earthenware or confectionary. They all come under the collective term of 'general cargo'.

The Chief Officer is usually that person designated on board the vessel who is responsible for the handling and safe stowage of all cargoes loaded aboard the ship. He is responsible for receiving the cargoes, and making sure that the holds are clean and ready to accept stowage and shipping in a safe manner. He is ultimately responsible for the carriage ventilation and delivery in good condition of all of the vessel's cargo.

In order to carry goods safely, the vessel must be seaworthy and the cargo spaces must be in such a condition as not to damage cargo parcels by ships sweat, taint or cause any other harmful factor. To this end the Chief Officer would cause a cargo plan to be constructed to ensure that separation of cargoes are easily identifiable and that no contamination of products could take place during the course of the voyage. The Chief Officer's prime areas of duty lie with the well-being and stability of the vessel together with the safe carriage of the cargo. Clearly, with the excessive weights involved with cargo parcels, the positive stability affecting the vessel's safe voyage could be impaired.

A correct order of loading with the capability of an effective discharge, often to several ports, must be achieved to comply with the safe execution of the voyage and also to stay within regulatory conditions, i.e. loadline requirements. This chapter is directly related to the details affecting stowage of particular cargoes and the associated idiosyncrasies, affecting the correct stow and carriage requirements to permit a lawful and successful venture.

Preparation of cargo spaces

The Chief Officer is generally that person aboard who is responsible for the preparation of the ship's holds to receive cargo. The preparations of the cargo compartments will usually be the same for all non-containerized general cargo parcels with additional specific items being carried out for specialized cargoes.

1. Holds and tween deck spaces should be thoroughly swept down to remove all traces of the previous cargo. The amount of cleaning will depend on the type of the previous cargo and the nature of the next cargo to be carried. On occasions the hold will need to be washed (salt water wash) in order to remove heavy dust or glutinous residues, but the hold is only washed after the sweepings and wastes have been removed.
2. Bilge bays and suctions should be cleaned out and tested, while the hold is being swept down. Tween deck scuppers should also be tested and rose boxes should be sighted and clear. All non-return valves in the bilge lines should be seen to be free and operating normally.

> ***Note:*** *If the previous cargo was a bulk cargo, then any plugs at the bilge deck angle should be removed to allow correct drainage.*

3. Check that all limber boards or bilge bay covers are in good condition and provide a snug fit. If bilges are contaminated, say from the previous cargo, and have noticeable odours, these should be sweetened and disinfected.
4. The spar ceiling (sometimes referred to as cargo battens) should be examined and replaced where necessary. In specific cases, like with an intended 'coal cargo', the spar ceiling should be totally removed from the compartment prior to loading.
5. All tween deck hatch coverings should be inspected for overall general condition and correct fitting. Tween deck guard rails, chains and stanchions should be fitted and seen to be in a good secure order.
6. Any soiled dunnage should be removed and, if appropriate, clean dunnage laid to suit the intended cargo to be loaded.
7. Checks should be made on the hold lighting, fan machinery, ventilation systems and the total flood fire detection/operation aspects.
8. Conduct a final inspection to ensure that the hold is ready to load. Some cargoes, like foodstuffs, may require the compartments to be inspected by a surveyor, prior to commencement of loading.

Duties of the Junior Cargo Officer (dry cargo vessels)

Cargo Officers will have a variety of duties before, during and after cargo operations begin. He/she should be aware that monitoring the cargo movements and ensuring parcels remain in good condition is protecting the owners' interests. Extensive ship keeping activities also go along with loading and discharging the vessel's cargo.

Prior to cargo operation

1. Check that the designated compartments are clean and ready to receive cargo.
2. Check that the drainage and bilge suctions are working effectively.
3. Ensure that cargo battens (spar ceiling) is in position and not damaged (some cargoes require cargo battens to be removed).
4. Make sure the relevant hatch covers are open and properly secured in the stowed position.
5. Check the rigging of derricks and/or the cranes are operating correctly.
6. Check that the hatch lighting's are in good order.
7. Order engineers to bring power to deck winches.
8. Inspect all lifting appliances to ensure that they are in good order.
9. Inspect and ensure all means of access to the compartments are safe.
10. Guard rails and safety barriers should be seen to be in place.
11. Ensure all necessary fire-fighting arrangements are in place.
12. Check that the ship's moorings are taught.
13. Note the draughts fore and aft.
14. Check that the gangway is rigged in a safe aspect.

During cargo transfer

1. Note all starting and stopping times of cargo operations for reference into the log book.
2. Note the movement of cargo parcels into respective compartments for entry onto the stowage plan.
3. Refuse damaged cargo and inform the Chief Officer of the action.
4. Monitor the weather conditions throughout operations.
5. Note any damage to the ship or the cargo-handling gear and inform the Chief Officer accordingly.
6. Maintain a security watch on all cargo parcels and prevent pilferage.
7. Tally in all special and valuable cargoes and provide lock-up stow if required.
8. Maintain an effective watch on the gangway and the vessels moorings.
9. Ensure that appropriate dunnage, separation and securing of cargo takes place.
10. Monitor all fire prevention measures.
11. Check the movement of passengers' baggage (passenger-carrying vessels).
12. Make sure all hazardous or dangerous cargoes have correct documentation and are given correct stowage relevant to their class (International Maritime Dangerous Goods (IMDG) Code).
13. Inspect compartments and the transit warehouse at regular intervals to ensure cargo movement is regular.
14. Inform Chief Officer prior to loading heavy lifts.
15. After discharge operations, search the space to prevent parcels being overcarried.

16. Ensure that the local by-laws are adhered to, throughout.
17. Note the draughts on the completion of loading/discharging.

After cargo operations

1. Close up hatches and lock and secure access points.
2. Inform engineering department to shut down power to deck winches.
3. Secure all lifting appliances against potential damage or misuse.
4. Enter the days working notes into the deck log book.
5. Inform the Chief Officer that the deck is secure and the current draughts.

Miscellaneous

The Chief Officer would ensure that the cargo stowage plan is kept updated when the vessel is in a loading situation. The officer's workbooks and tally sheets would be used at this stage. He would also at some time order the density of the dock water to be ascertained by means of the hydrometer. Cargo loaded/discharged, being then ascertained by means of the deadweight scale.

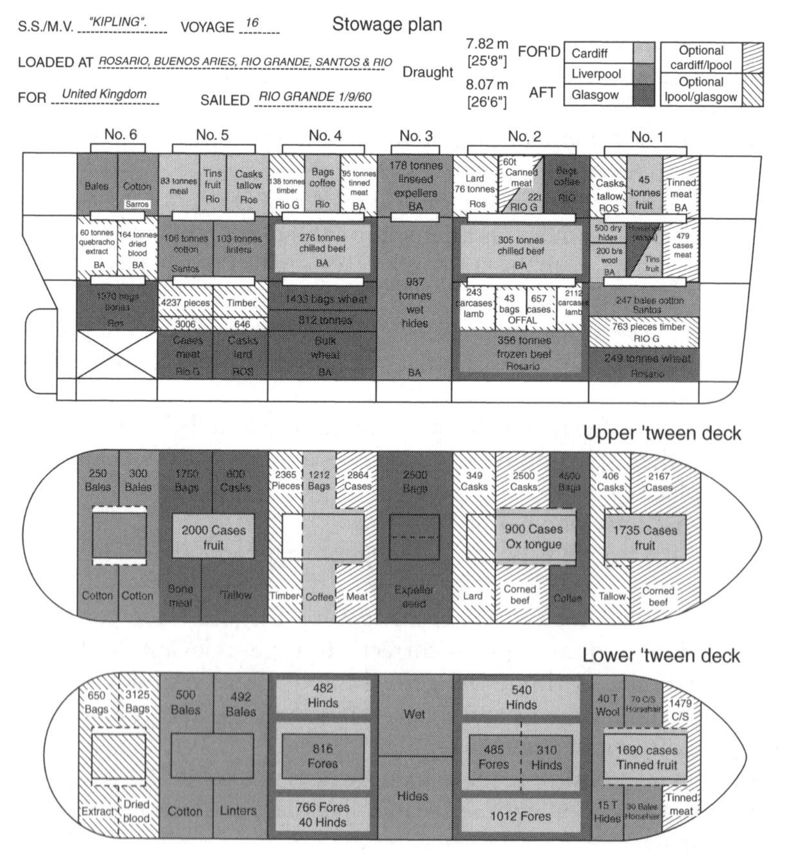

Fig. 3.1 Example of open stow, general cargo plan.

The cargo stowage plan

The function of the 'stowage plan' is to identify the various cargo parcels by quantity, destination and nature of goods (Figure 3.1).

It permits the Chief Officer to assess the number of stevedore gangs for respective compartments and the times associated with cargo operations. Additionally, it shows the location of special cargoes like 'heavy lifts' or 'hazardous goods', valuables and the lock-up stow goods.

Ventilation and fire-fighting procedures can be influenced by the disposition or respective cargoes, while the owners/Charterers are provided with notification of available space between discharge ports, useful for diverting the vessel for further cargo.

Stowage plans provide the following relevant details in addition to the pictorial cargo distribution plan:

Cargo distribution summary (tonnes)

Port of discharge	Colour code	No. 1 Hold	No. 2 Hold	No. 3 Hold	No. 4 Hold	On deck	Port total
1st							
2nd							
3rd							
4th							
5th							
6th							
Total							

Deadweight particulars

Draughts:
Forward ---
Aft ---
Mean ---
Density correction ----------------------------------- Scale D/W
S.W. draught ----------------------------------- tonnes

Cargo ---------------------------
Fuel ---------------------------
Fresh water ---------------------------
Ballast ---------------------------
Stores ---------------------------

Total D/weight =
Scale D/weight =

Difference
________________________ Tonnes __________

The above information with the ship's name and port(s) of loading, together with date of sailing, are all included on the plan. Fuel, ballast and fresh water are usually depicted in alternative colours to colour codes as used for discharge ports.

Tanker stowage plan (profile + pipeline) (Chapter 5)

Particularly useful with product tankers where the disposition of grades of cargo can be clearly illustrated. The plan can ensure that adjacent tanks are not likely to generate contamination.

The pipeline system is often employed in conjunction with the plan to ensure that correct lines are operational with the correct grade of product. Quantities and type of each product can be easily identified, but clearly this plan is not as detailed as with say an open stow general cargo vessel carrying many different types of cargoes.

Roll-on, Roll-off stowage plans (plan view) (Chapter 7)

These are generally computer generated and like other stowage plans helps to identify individual units. This is specifically required for any units carrying dangerous/hazardous products. It also permits the order of discharge to be pre-arranged. Modern loadicators are usually involved with the planning of cargo stowage with Roll-on, Roll-off (Ro-Ro) vessels. They permit known weights aboard the vessel to be pre-programmed and the centre of gravity of each unit, with its respective stowage space, can be entered to provide the ships overall metacentric height (GM), very quickly.

Container stowage plans (elevation + cross section) (Chapter 8)

Container stowage plans are a proven way of tracking specific units during the sea passage. The plan identifies each unit and allows shippers to estimate arrival times and the whereabouts of their goods during every stage of shipment. It is also an effective security aspect for knowing which unit is where and tracing what goods are in what particular unit.

Pre-load plans

Provides a provisional distribution plan for the intended cargo parcels. They may be accompanied by capacity space set against cargo capacity to reflect unused space. They can determine access points or detail pipeline arrangements prior to commencing cargo operations. They are generally used on all types of vessels.

Steel cargoes

Probably the most physically dangerous, of all cargoes, is steelwork. Steel cargoes tend to come in all shapes and sizes, from the biggest 'casting' to the long steel 'H' girders used extensively in the construction industry. Long and heavy loads are difficult to control and the slightest contact with surrounding structures could generate extensive damage or injury to personnel. The fact that in most cases they are rigid and heavy, makes handling safely extremely difficult. One of the exceptions to this, amongst the steel cargoes, is the loading and carriage of bulk scrap metal. Also a dangerous cargo but for different reasons. It is loaded/discharged by heavy grabs that generally cause some fall out between quayside and shipside.

Steel coils

Another form of steel cargo is heavy steel coils. The round shape makes this cargo a high risk to shifting in a seaway, especially if it is not properly secured. In the event of the cargo shifting, the ship could expect to take on a list which, if considered dangerous, could necessitate the vessel altering course to a port of refuge. The prime purpose for this would be to discharge and then to re-secure cargo, a costly business.

Steel coils are normally stowed in a double tier with the bottom coils on athwartships dunnage and wedged against athwartships movement, each coil being hard-up against the next (Figures 3.2 and 3.3). The objective is to

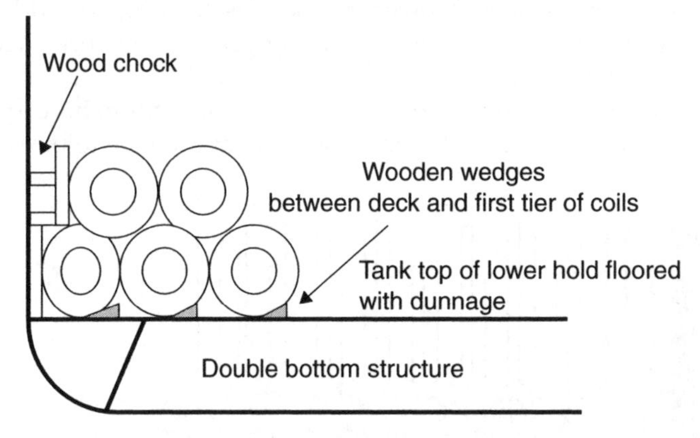

Fig. 3.2 Secure stowage of steel coils.

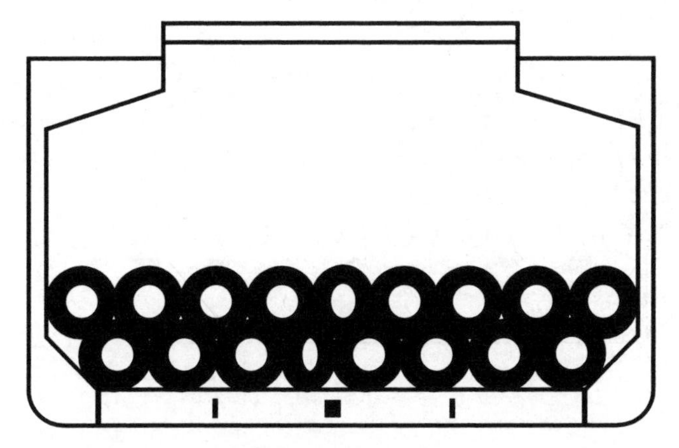

Fig. 3.3 Steel coil loading Diamond bulker design with complete double hull is able to stow two tiers of heavy steel coils (each up to 25 tonnes) across the hatch.

form a large immovable stow with any void spaces between coils chocked off with dunnage. End of stows would be fenced with timber battens and 'locking coils' together with the top tier of coils would most certainly be lashed with steel wire rope lashings. With such a heavy cargo, the ship could be expected to reach her loadline marks quickly leaving some considerable broken stowage with this type of cargo.

The turn of the bilge is protected by vertical dunnage and the second tier of coils is then placed on top. As the second tier is filled, it should be recognized that the stow will have 'key' locking coils and these should be lashed into position by steel wire lashings, while remaining accessible. It is also worth noting that the size and weight of individual coils is not always uniform and, as such, differences create small gaps between the cargo stow. These gaps, where substantial, should be 'chocked' with baulks of timber, while wedges can be used to prevent movement between smaller gaps (Figure 3.4).

Cargo Officers should be wary when working this cargo as the method of lifting during the loading process will usually be by a standard crane with adequate safe working load. However, when discharging, the odd coil may be of a heavier variety and cause the lifting gear to be overloaded (some coils go up to 10 tonnes).

The main concern for any Ship's Master with coils within his cargo is that they are correctly secured and to this end it is not unusual to hire a rigging

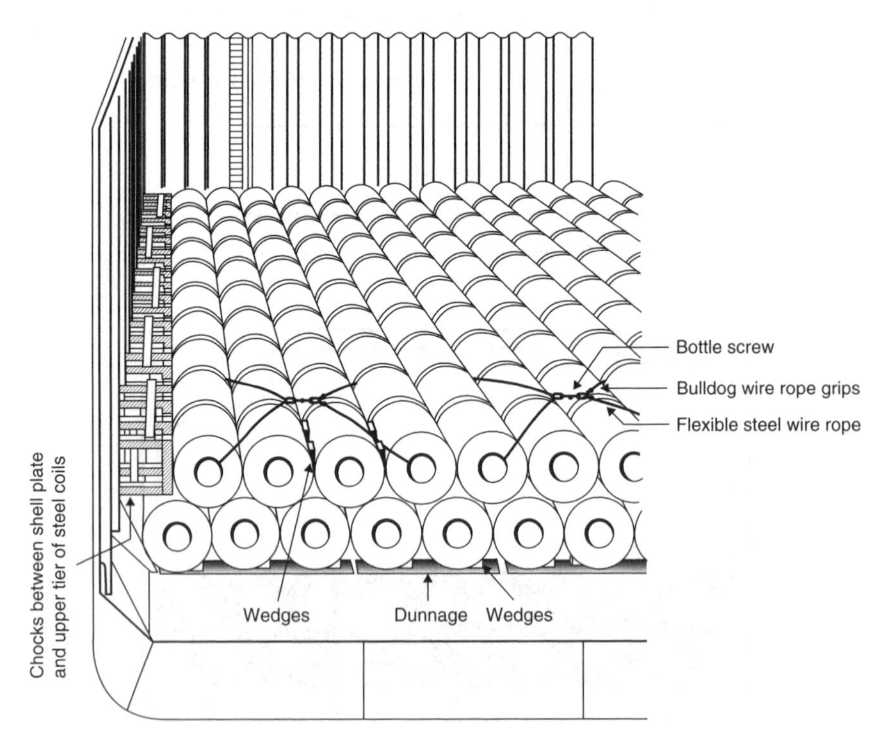

Fig. 3.4 Securing of steel coils.

gang during the period of loading. The ship's stability must also be taken into consideration with such a heavy cargo and as such, coils tend to be always loaded in lower holds, as opposed to tween deck spaces. Such a loading pattern would tend to generate a favourable 'GM'.

The quantity of cargo is usually restrictive, because its overall weight will soon bring the vessel to her loadline marks. Geographically, the loading port will dictate which loadline zones the vessel must route through to reach her port of discharge and as such, the passage plan should reflect a route that would minimize the ship's rolling pattern wherever possible.

Steel plate

Steel plates come in a variety of sizes from very long to heavy bundles of about 2 m in length. Various methods are employed to handle this commodity from plate clamps on chain slings to electromagnetic expanding beams (Figure 3.5). The weight and overall size tend to make this an awkward and dangerous cargo to load or discharge and once stowed requires chain securings.

Being heavy it is usually given bottom stow and floored with dunnage to accommodate any overstowing. Large plates may incur damage to the vessel and/or other cargoes during movement and careful handling should be the order of the day.

Modern-handling techniques where steel plate is a regular cargo, tend to employ gantry cranes working with electromagnetic, expandable beams. These are similar to steel stockyard cranes and are now seen in Port Terminals working in a similar manner to container gantry cranes, e.g. at the Port of Immingham, UK.

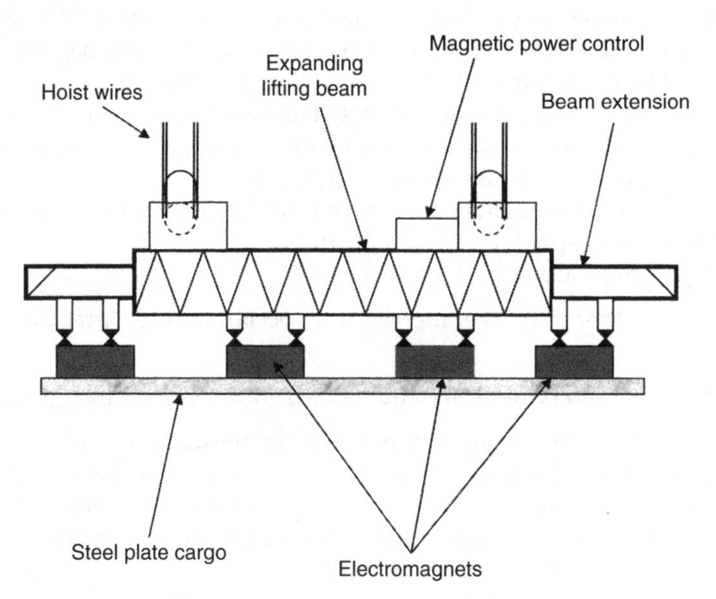

Fig. 3.5 Operating principle of Telescopic Magnetic Lifting Beam.

Bagged cargoes

There are many examples of bagged cargoes: fishmeal, grain, beans, cocoa, etc. to name but a few. They may be packed in paper bags like cement, or Hessian sacks, as employed for grain or bean products, loading taking place either in containers or on pallet slings. Size of bags tends to vary depending on the product, and are seen as a regular type of package for general cargo vessels.

However, handling bagged cargo is expensive by today's standards and many of the products lend more easily and more economically to bulk carriage or container stow. Where bags are stowed they should be on double dunnage, stacked either bag on bag or stowed half bag as shown in Figure 3.6.

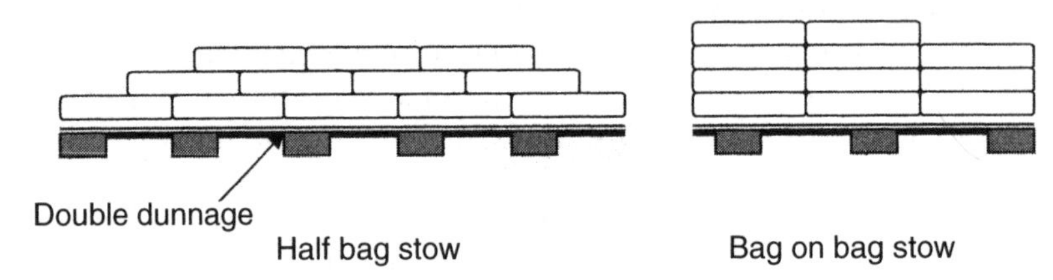

Fig. 3.6 Examples of bagged stowage.

When receiving bagged cargo the bags should be seen to be clean and not torn. Neither should they be bled in order to get a few extra bags into the compartment. Such an action would only increase the sweepings after discharge and lead to increased cargo claims.

Slings should be made up, in or close to, the square of the hatch. If they are made up in the wings, then bags are liable to tear as the load is dragged to the centre. Stevedores should not use hooks with paper bags and bags should not be hoisted directly by hooked lifting appliances.

Shippers frequently provide additional unused bags to allow for residual sweepings. This allows for all bags being discharged ashore, even torn bags, to ensure that a complete tally is achieved.

Bags containing oil seeds of any type must be stowed in a cool place as these are liable to spontaneous combustion.

Examples of products for bag stowage:

Bone meal – other than keeping dry, no special stowage precautions are required.

Cattle food – should be kept dry and away from strong smelling goods.

Cement – paper bags require care in handling. Stow in a dry place and not more than fifteen (15) bags high. Alternative carriage in bulk in specially designed ships for the task. Bilges should be rendered sift proof and compartments must be thoroughly clean to avoid contamination which would render cement useless as a binding agent.

Chemicals – prior to loading check the IMDG Code and provide suitable stow.

Cocoa – stow away from heat and from other cargoes which are liable to taint.

Coffee – requires plenty of ventilation and susceptible to damage from strong smelling goods.

Copra – dried coconut flesh. Liable to heat and spontaneous combustion. It could taint other cargoes and cause oxygen deficiency in the compartment. Requires good surface ventilation.

Dried blood – used as a fertilizer and must be stowed away from any cargoes liable to taint (similar stow for bones).

Expeller seed – must be shipped dry. It is extremely high risk to spontaneous combustion and must not be stowed close to bulkheads, especially hot bulkheads.

Fishmeal – gives off an offensive odour and requires good ventilation. This cargo is liable to spontaneous combustion and requires continuous monitoring of bags and surrounding air temperatures. Bags should not be loaded in a wet or damp condition, or if they are over 35°C or + 5°C above ambient temperature whichever is higher.

Flour – easily tainted. The stow must be kept dry and clear of smelly goods.

Potatoes – loaded in paper sacks. Require a cool, well-ventilated stow.

Quebracho extract – this is a resin extract used in the tanning industry. Bags are known to stick together and should be separated on loading by wood shavings.

Rice – see next page.

Salt – requires a dry stowage area.

Soda ash – should be stowed away from ironwork and foodstuffs, and must be kept dry.

Sugar – also carried as bulk cargo. Bagged green sugar exudes a lot of syrup. Stowage should be kept clear of the ship's side as the bags are susceptible to tearing as the cargo settles. Dry refined sugar and wet or green sugar must not be stowed together. Cover steelwork with brown paper for bulk sugar and keep dry.

Rice

Rice is considered as a 'grain' cargo and would need to meet the requirements of the Grain Regulations affecting stowage. A ship's condition format would be required to show the cargo distribution and a curve of statical stability for the condition would need to be constructed.

Rice cargoes are now usually carried in bulk. This eliminates the costs of handling bags for the shipping phase. It is more economical and common to bag rice products at the distribution stage.

Rice contains a considerable amount of water and is liable to sweat. It must be well ventilated and not allowed to become moist or it will start to rot and give off a pungent smell which could affect other rice cargoes in the vicinity. It is also known to give off carbonic acid gas (a weak acid formed when carbon dioxide (CO_2) is dissolved in water).

Ventilators should generally be trimmed back to wind, although matured grain rice will require less ventilation than new grain rice. In any event, a void space between the deck head of the compartment and the surface of the stow should be left bearing in mind the possibility of cargo movement and the necessity to employ shifting boards. Surface ventilation should be ongoing to remove warm air currents rising from the bulk stow.

Prior to loading rice, the compartments should be thoroughly cleaned, bilges sweetened and made sift free. A lime coating is recommended, together with a cement wash. Their condition must be such to pass survey inspection. The hold ceiling should be stain free and covered by a tarpaulin or separation cloth. To this end an adequate supply of matting and separation cloths are to be recommended.

If compartments are only partly filled, then bagged rice with suitable separation cloths may be used to secure the stow. Bags for rice are usually of a breathable man-made, interwoven fabric. A ship loading rice would need a Certificate of Authorization, or alternatively the master would need to show that the vessel can comply with the carriage regulations to the satisfaction of an Maritime and Coastguard Agency (MCA) Surveyor.

Modern loading methods usually employ chutes, while pneumatic suction systems are often engaged for the discharge process. Working capacity of distribution and suction units is up to about 15 000 tonne/h (stowage factor for rice in bags = 1.39 m^3/tonne, or bulk stow = 1.20 m^3/tonnes). *Note*: See additional reference in Chapter 4.

Bale goods

Various types of goods are carried in bales, either in open stow or containerized. Bales in open stow are normally laid on thick single dunnage of at least 50 mm in depth. Bales are expected to be clean with all bands intact. Any stained or oil marked bales should be rejected at the time of loading. All bales should be protected against ships sweat and the upper level of cargo should be covered with matting or waterproof paper to prevent moisture from the deck head dripping onto the cargo surface.

Examples of bale cargoes:

Carpets – a valuable cargo which must be kept dry. Hooks should not be used. More commonly carried in containers these days.

Cotton/cotton waste – bales of cotton are highly inflammable and stringent fire precautions should be adopted when loading this cargo. A strict no-smoking policy should be observed. If the bales have been in contact with oil or are damp they are liable to the effects of spontaneous combustion. Generally, a dry stowage area is recommended.

Esparto grass – these and products like hay and straw bales are high risk to spontaneous combustion especially if wet and loosely packed. Poorly compressed bales should be rejected. If carried on deck these bales should be covered by tarpaulins, or other protective coverage.

Fibres – such as jute, hemp, sisal, coir, flax or kapok are all easily combustible. A strict no-smoking policy should be observed at all stages of contact. Bales must be kept away from oil and should not be stowed in the same compartment as coal or other inflammable substances or other cargoes liable to spontaneous combustion.

Oakum – this is hemp fibres impregnated with pine tar or pitch. It is highly inflammable and strict no-smoking procedures should be adopted. It is also liable to spontaneous combustion.

Rubber – if packed in bales these give an unstable platform on which to overstow other cargoes, other than more bales of rubber. Crêpe rubber tends to become compressed and sticks to adjacent bales and talcum powder should be dusted over the bales to prevent this stickiness between bales. Polythene sheeting with ventilation holes is also used and is now in more common use for the same purpose. Up-to-date methods tend to wrap the whole bales separately in polythene to eliminate the sticking element.

Tobacco – usually stowed in bales in open stow. It is liable to taint other cargoes and is also susceptible to taint from other cargoes in close proximity. The stowage compartment should be dry and kept well ventilated or there is a risk of mildew forming.

Wood pulp – must be kept dry. If it is allowed to get wet it will swell and could cause serious damage to the steel boundaries of the compartment. Notice metacentre (M) 1051 recommends that care should be taken to ensure that no water is allowed to enter the compartment. To this end all air pipes and ventilators should be sealed against the accidental ingress of water.

Wool – can be shipped in either scoured or unscoured condition. The two types should not be stowed together. Bales should be well dunnaged and provided with good ventilation. Slipe and pie wools are liable to spontaneous combustion and should, if possible, be stowed in accessible parts of the hold.

Loading, stowage and identification of cargo parcels

In order to ensure correct handling and stowage of goods, cargoes tend to be labelled and marked with instructions on the side of respective packages (Figure 3.7). Cargo is shipped all over the globe and not all countries of discharge are English speaking – to this end labels are of a pictorial display. Cargo Officers can monitor from the labelling that instructions are complied with and that stowage practice is as per shippers' instructions.

Fig. 3.7 Example labels provide instructions to stevedores for correct stowage practice.

Stowage of wine

Wine was often carried in barrels, and in some cases still is. However, bulk road tankers, and even designated wine carriers, are engaged in the shipment of large quantities of wine in bulk. Where barrels are transported they should be stowed on the side (bilge), with the 'bung' uppermost (Figure 3.8(a)). The stow should not be greater than eight (8) high and the first height level should be laid on a bed to keep the bilge free. 'Quoins', a type of wedge arrangement, are used to support the barrels and prevent them from moving (Figure 3.8(b)).

Barrels are heavy, with a capacity of 36 imperial gallons (164 l) and normally require two men to handle and stow in a fore and aft direction. Modern aluminium casks have, to some extent, replaced the old wooden barrels but some companies still use the old-fashioned wood barrels for their product.

Barrels are given underdeck stowage and would not generally be taken as deck cargo.

Where wine is not shipped in bulk-holding tanks or barrels the more popular method in this day and age is to pre-bottle the commodity and export in cartons usually in a container. Distinct advantages are associated with this method, in that pilferage is reduced with the bottled wine under lock and key. Containers are easily packed and sealed under customs controls. Mixed commodities, like spirits or beer, can also be packed into the same container. Once sealed, transport and shipping via a container terminal is usually trouble free.

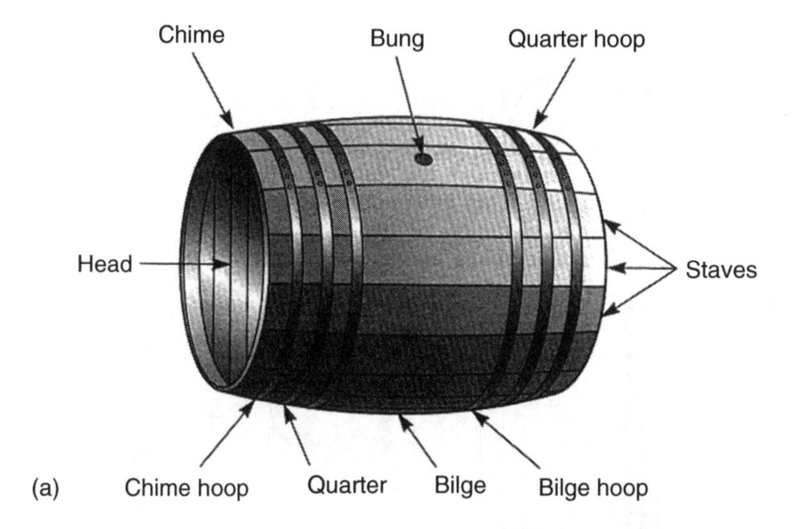

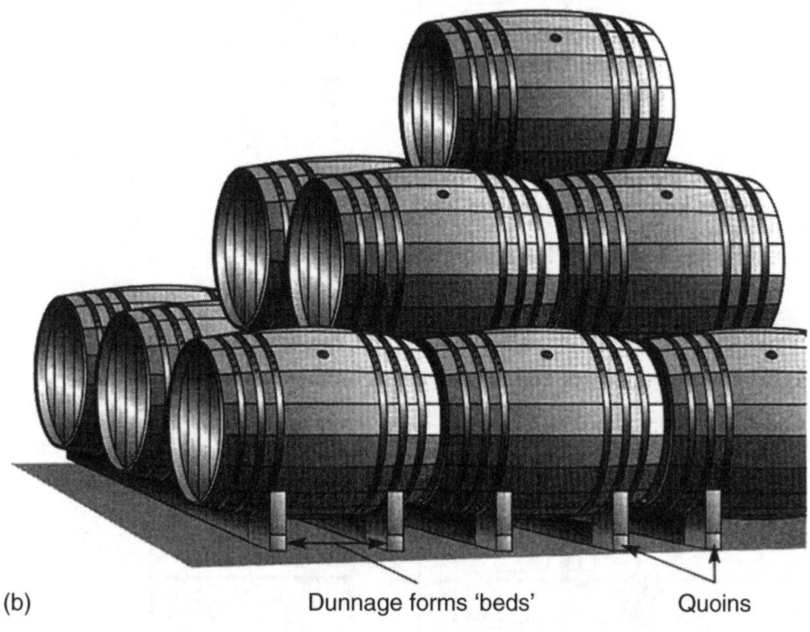

Fig. 3.8 Wine barrels.

Barrels are used more these days to allow wines to mature, rather than as transport vessels. They are awkward to handle and have difficulties in stowage. The art of the 'cooper' is also becoming scarce and if barrels are damaged in transit it becomes expensive to effect repairs.

Occasionally, barrels are still employed but with specialist commodities or shipped from one wine cellar to another where surplus casks are available or required.

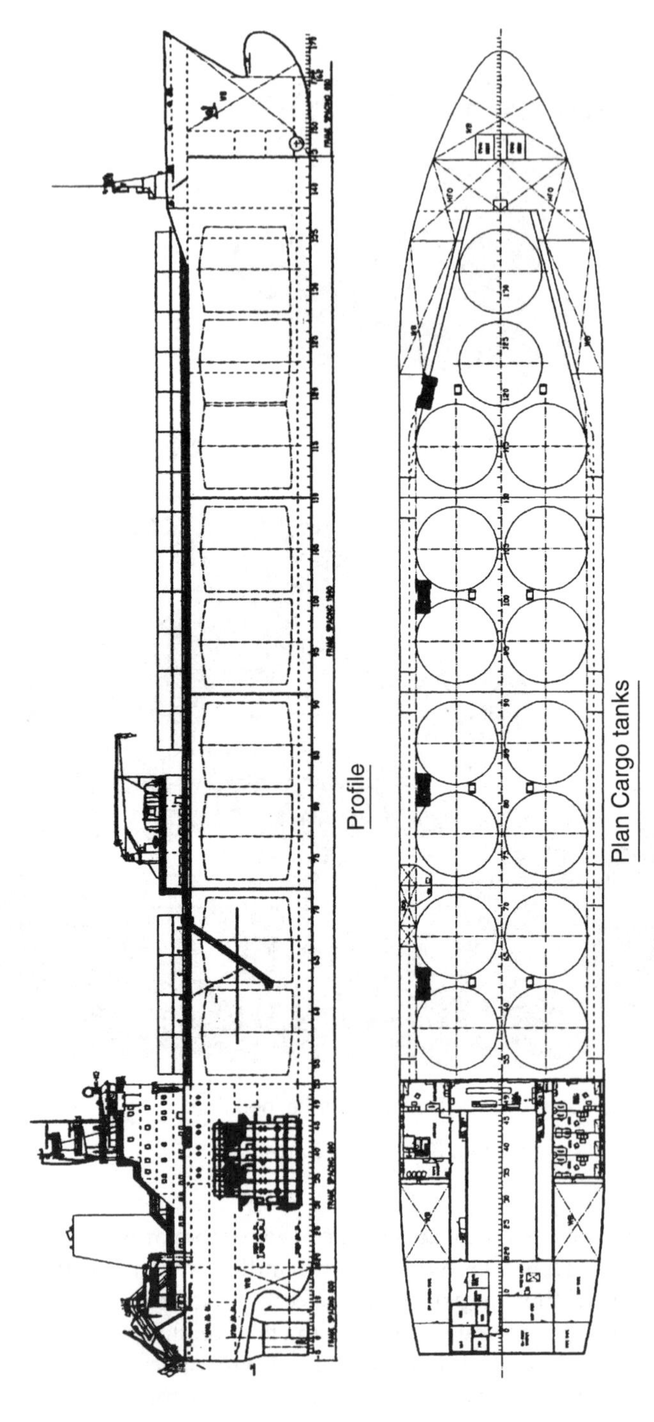

Fig. 3.9 Profile and tank disposition of 'Carlos Fischer'. Reproduced with kind permission from Motor Ship.

Bulk fluid products

Some products like wine and fruit juices have generated the construction of specialized transports, specifically for the carriage. An example of this is seen with the 'Carlos Fischer' fitted with free standing, stainless steel tanks for the purpose of shipping bulk 'orange juice' (Figure 3.9).

The ship is 42 500 dwt (deadweight tonnage), and is engaged in shipping concentrated orange juice from estates in Brazil. It is double hulled but not classed as a tanker, having four holds each with four vertical cylindrical fruit juice tanks. Cargo piping running through the holds is led to manifolds in lockers in the deckhouse.

Case goods

Case goods lend particularly to a general cargo open stow but can be containerized depending on size. Heavy cases should always be given bottom stow with the lighter cases on top. If the contents of the case are pilferable, then they should be loaded into a lock-up stow and tallied in and tallied out.

Slinging of case goods will be directly related to their weight and may be fitted with identified lifting points. Care should be taken that such lifting points are attached to the load and not just to the package (Figure 3.10).

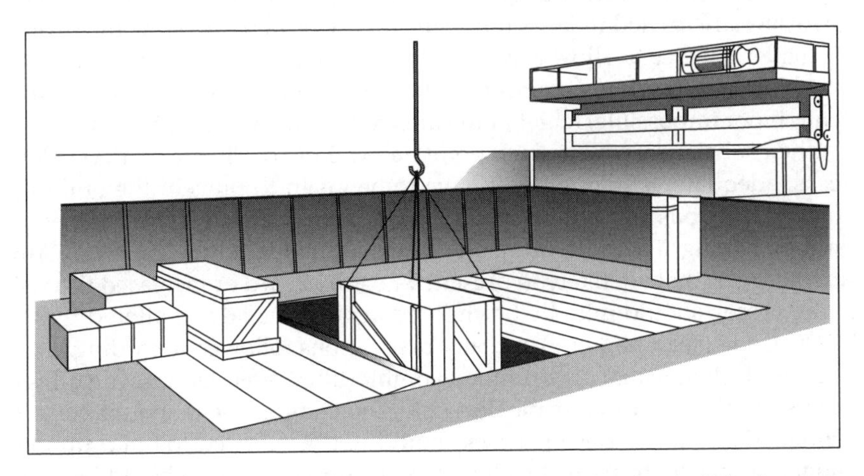

Fig. 3.10 Example of case goods/general cargo being loaded/discharged.

Specific case goods, i.e. glass, may have special stowage requirements. This would probably be marked as 'Fragile' or 'This way up' and require side, end on stowage. Crated cars or boats would expect to be loaded on level ground, and generally other crated goods would be treated as case goods depending on the nature of the contents.

Fork lift truck operations are often employed with the movement and stowage of heavy case goods both in the warehouse, on the quayside and aboard the vessel. However, the use of fork lift trucks inside the hold tends to be restrictive with case goods because they are so bulky. The fork lift truck needs open deck space to allow manoeuvring and as large cases quickly start to fill the manoeuvring space, landing becomes the only method to continue loading.

The loading and carriage of drums

Cargo in drums is not unusual and can be varied by way of chemicals, oils, paints, dyes, even sheep dip. Drums may differ in size, but a 50-gallon drum is probably the most common size for oils and is often used for own ship's stores of lubricating or diesel oil.

They are often taken as deck stow. In such an event, they would be protected by nets or a timber built compound to keep the stow tight, depending on the number of drums carried. Where upper decks are covered, this may necessitate a catwalk being built over the drums in order to provide accessibility to all parts of the vessel when at sea. In any event, drum cargoes are placed on single dunnage and are invariably secured by wire lashings, with or without nets, to prevent movement of the cargo when at sea.

Concern with such cargo may arise with the obvious problem of a leaking drum. If such an occurrence did take place, the action would depend on the contents of the drum, the associated effects on other cargoes, the potential fire risk and the ability to get at the affected drum(s). To this end, where corrosives are carried in large numbers, it may be better to stow the drums in smaller batches to allow accessibility to damaged units, as opposed to a total block stow of many drums together. Such a block stow may prove difficult, if not impossible, to get at the affected drums when in transit.

When substances with a flash point below 23°C (73°F) are carried below decks, adequate ventilation will have to be given to prevent the build up of any dangerous concentrations of inflammable vapours. Low flash point cargoes having a wide flammable range are extremely hazardous. Any such cargoes, that are likely to present a health hazard or increased fire risk to the vessel, should initially be checked against the advice offered by the IMDG Code (see Chapter 9) and any precautions followed accordingly.

Underdeck stowage of drummed commodities often tend to run a high fire risk with or without explosion risk. The compartment should be well ventilated and any gases or fumes should not be allowed to build up into dangerous concentrations. Prudent use of cargo hold fans should be exercised while on the voyage to ensure a continued safe atmosphere within the compartment and a no-smoking policy must be observed at all times.

Casks are manufactured in aluminium and are used extensively for 'beer'. They are comparatively light and may be full or empty. They require a compact stow and are often netted to prevent movement when in open stow.

A cargo of ingots – Copper, lead or tin ingots are all very heavy concentrated cargo parcels and require bottom stow, similar to the iron cargoes of castings, iron billets and long steelwork.

Lighter goods may be stowed on top of ingots but a secure separation between cargoes is desired. Ingots cannot be stowed high and are difficult to work on top of the cargo without a dunnage floor. Ingots are often baled and banded, but are sometimes shipped as single-bar elements being floor stacked. Ingots can be considered a valuable cargo and are usually tallied in and tallied out at discharge.

Cable reels – large wooden reels with power cable rove around a central core are carried as general cargo. They are stowed in the upright position, on a firm deck and should be secured against any pitching or rolling of the vessel when in a seaway. They can be quite, large, 3–4 m in diameter, and consequently may be considered as a heavy load, especially if the cable contains a steel construction element.

Designated 'Cable Ships' with telegraph cable tend to load the cable directly into specially constructed cylindrical tanks in specialized cable holds. Such cables should not be confused with the Cable Reels discussed as general cargo.

Paper cargoes – paper may be carried in many forms from waste paper to newsprint. The compartment, in whatever form the paper is to be carried, must be in a dry condition and well ventilated. Newsprint is carried in rolls which are normally stowed on their ends to avoid distortion, preferably on double dunnage.

A ship's steelwork would normally be protected with waterproof paper to prevent ships sweat from damaging the rolls. Hooks should not be used during the loading or discharge periods. On occasions, like in tween decks, the rolls may be stowed on their sides. If this is done, they should be chocked off to prevent friction burns and movement when the vessel is at sea.

Rolls of paper should be sighted as being unmarked by oil or other similar stains on loading. Once on board, the cargo should be kept clean and not allowed to become contaminated by any form of oil or water.

Dried fruits – these include: apricots, currents, dates, figs, prunes, raisins and sultanas. May be shipped in cases, cartons, small boxes or even baskets. However carried, they must be stowed away from cargoes which are liable to taint. Dried fruits tend to give off a strong smell and generally may contain drugs and insects which could contaminate other cargoes, especially foodstuffs. The fruit itself is liable to taint from other strong odorous cargoes and stowage should be kept separate in cool well-ventilated compartments. Tween deck stowage is preferred, but if stowed in lower holds adequate ventilation must be available throughout the course of the voyage. If in open stow, good layers of dunnage are recommended to assist air flow and the cargo should not be overstowed.

Garlic and onions – shipped in bags, cases or crates and these give off a pungent odour and must be stowed clear of other cargoes liable to taint. It is essential that onions and garlic are provided with good ventilation, similar to fresh fruit. Considerable moisture will be given off onions and adequate drainage facilities would be expected.

Fresh fruit – apples, apricots, pears, peaches, grapefruit, grapes, lemons and oranges can be carried quite successfully in non-refrigerated compartments, the proviso being that adequate dunnage is used along with good ventilation. In the event that mechanical ventilation is not used then hatches should be opened (weather permitting). Fruit, especially green fruit, gives off a lot of gas and extreme care should be exercised before entering any compartment stowed with fresh fruit. Following the discharge of fruit the holds should be well aired and deodorized.

Cargo monitoring and tallying

Tallying – all cargoes are tallied on board the vessel and for monitoring the cargo parcels in this manner, specialized 'tally-clerks' are employed. These clerks tend to reflect the shipper's interest, while others so engaged by the ship may represent the owner's or ship's operator's interests. Cargo parcels are not only tallied into the ship but also tallied out at the port of discharge.

Tally counts are important, especially in the case of valuable effects, or short quantities being delivered to the ship. Cargo claims draw on tally information to substantiate quality and quantity as and when disputes evolve between the ship and the shipper, bearing in mind that the ship's personnel are there to protect the shipowner's or Charterer's interest.

Mate's Receipts tend to be the supporting document which denotes the quantity, marks, description and the apparent condition of goods received on board. It is usually signed by the Ship's Chief Officer, hence the name 'Mate's Receipt'.

CARGO shipped on Board "________________"

In good condition excepting where otherwise stated

*Port of Shipment*________________ *Date*________________

*Destination*________________ *Hatch No*________________ *ex*________________

NB Ships Tally Clerk to record all visible damage

MARKS	PACKAGES	SEPARATE NUMBERS	TOTAL

Ships tally Clerk

____

Fig. 3.11 Tally clerk's account.

It is important that the details of the cargo are correctly stated on the Mate's Receipts as it is from these that the 'Bills of Lading' (B/L) could be prepared. The Bills of Lading are sent to the various consignees, who will in turn present them to the master before the cargo is handed over. The Bills of Lading are the consignee's title to the goods stated and he therefore can expect to receive those goods as described. In the event of the goods not being in the same condition as stated on the B/L, by way of quantity or quality, then the shipper could make a claim against the ship for any discrepancy.

Ship's Officers should bear in mind that they are temporary custodians of goods which belong to a third party. As such, they must endeavour to keep them in the same condition as that in which they were received aboard the vessel. As far as possible damaged cargo or damaged packages should be rejected for shipping.

Cargo sweat and ventilation

A great number of cargo claims are made for merchandise which has been damaged in transit. Much of this damage is caused by either 'ships sweat' or 'cargo sweat' and could be effectively reduced by prudent ventilation of cargo spaces.

Sweat is formed when water vapour in the air condenses out into water droplets once the air is cooled below its dew point. The water droplets may be deposited onto the ship's structure or onto the cargo. In the former, it is known as 'ships sweat' and this may run or subsequently drip onto the cargo. When the water droplets form on cargo this is known as 'cargo sweat' and will occur when the temperature of the cargo is cold and the incoming air is warm.

To avoid sweat and its damaging effects it is imperative that 'wet and dry' bulb temperatures of the air entering and the air contained within the cargo compartment, are taken at frequent intervals. If the temperatures of the external air is less than the dew point of the air already inside the space, sweating could well occur. Such conditions give rise to 'ships sweat' and is commonly found on voyages from warm climates towards colder destinations. Similarly, if the temperature of the air in the cargo compartment (or the cargo) is lower than the dew point of incoming air, sweating could again occur, giving rise to 'cargo sweat'. This would be expected on voyages from cold places towards destinations in warmer climates.

If cargo sweat is being experienced or likely to occur, ventilation from the outside air should be stopped until more favourable conditions are obtained. However, it should be noted that indiscriminate ventilation often does more harm than no ventilation whatsoever. It is also of concern that variation in the angles of ventilators away from the wind can cause very different rates of air flow within the compartment. The angle at which the ship's course makes with the wind also affects the general flow of air to cargo compartments. In general, the greatest air flow occurs when the lee ventilators are trimmed on the wind and the weather ventilators are trimmed away from the wind. This is known as through ventilation (Figure 3.12).

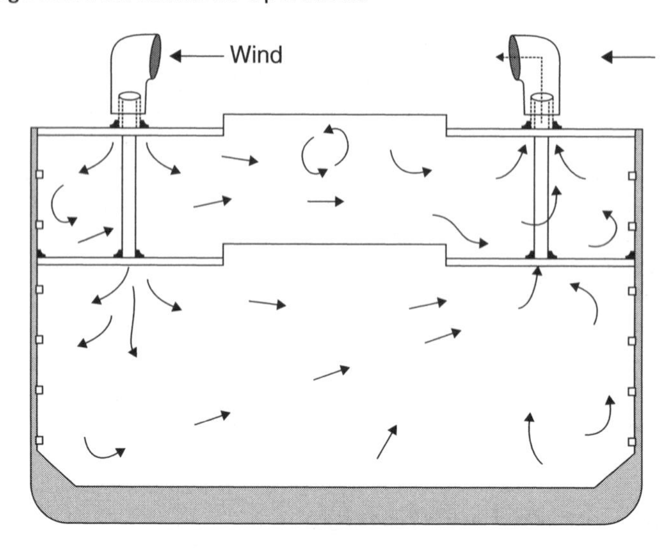

Fig. 3.12 Showing air circulation with lee ventilators on the wind and weather vents off. This is through ventilation.

Forced ventilation – if the dew point temperature in the cargo compartment can be retained below the temperature of the ships structure, i.e. decks, sides, bulkheads and the cargo, there would be no risk of sweat forming. Such a condition cannot always be achieved without some form of mechanical (forced) ventilation from fans or blowers.

There are several excellent systems on the commercial market which have the ability to circulate and dry the air inside the cargo holds. Systems vary but often employ 'baffle' plates fitted in the hold and tween decks so that air can be prevented from entering from the outside when conditions are unfavourable. Systems re-circulating the compartment's air can also operate in conjunction with dehumidifying equipment to achieve satisfactory conditions pertinent to relevant cargo.

Cargo battens (spar ceiling) – the purpose of the wooden cargo battens, which can be fitted horizontally or vertically, is to keep the cargo off the ship's inner steel hull. This arrangement produces an air gap of about 230 mm between the steelwork and the cargo surface, and subsequently reduces the risk of cargo sweat damaging cargo parcels. It is normal practice with some bulk cargoes, when carried in holds fitted with spar ceiling, to remove the wood battens to reduce the damage incurred to the wood, prior to loading, e.g. coal (Figure 3.13).

Dunnage – timber slats of a thickness of about 35/40 mm which are ordered in bundles by the Ship's Chief Officer. The purpose of dunnage, which can be laid either singularly or in a criss-cross double dunnage pattern, is to provide an air gap to the underside of the cargo. This allows ventilation around all sides of the cargo stow. This is again to effectively reduce the risk

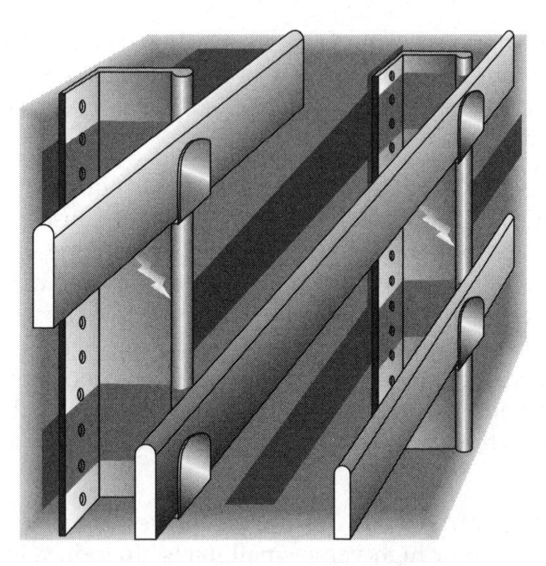

Fig. 3.13 Cargo battens fitted horizontally to allow separation of the cargo from the ship's inner steel hull.

of sweat damage to cargo. Dunnage should be in a clean condition and not oily or greasy as this could cause contamination to sensitive cargoes.

Tank top ceiling – a wood sheathing which covers the steelwork of the tank top, in way of the hatchway in the lower hold. This timber flooring not only protects the tank tops but also lends to a non-skid surface in the hold. It generally assists drainage of any moisture in the space and can be used in conjunction with a single-dunnage layer.

Contamination – cargoes which taint easily, e.g. tea, flour, tobacco, etc. should be kept well away from strong smelling cargoes. If a pungent cargo has been carried previously, i.e. cloves or cinnamon for example, the compartment should be deodorized before loading the next cargo.

Dirty cargoes should never be carried in the same compartment as clean cargoes. A general comparison of dirty cargoes would include such commodities as oils, paints or animal products, whereas clean cargoes would cover the likes of foodstuffs or fabrics. Obviously some notable exceptions in each of the two classes are to be found.

Separation of cargoes – it is often a requirement when separate parcels of the same cargo are carried together that a degree of separation between the units is essential. Depending on the type of goods being shipped will reflect the type of separation method employed. Examples of separation materials include colour wash, tarpaulins, burlap, paper sheeting, dunnage, chalk marks, rope yarns or polythene sheets.

The idea of separation is to ensure that the cargo parcels, although maybe looking the same, are not allowed to become inadvertently mixed.

Optional cargo – optional cargo is cargo which is destined for discharge at either one, two or even more ports. Consequently, it should be stowed in such a position as to be readily available for discharge, once the designated port is declared.

Overcarried cargo – if cargo meant for discharge is not discharged it is said to be overcarried to the next port. Such an event causes inconvenience, extra cost and additional paperwork. To this end hatches are searched on completion of discharge to ensure that all the designated cargo for the port of discharge has indeed left the ship – a method of checking against the cargo plan and the cargo manifest and comparing figures with the tally-clerks. It must be said, however, that this is not foolproof, especially if pressures are being applied to finish cargo operations and sail, and possibly departing before the holds have been properly examined for overcarried cargo pieces.

Pilferage – certain cargoes always attract thieves. Notable items include spirits, beer, tobacco or high value small items. To reduce losses such cargoes should be tallied in and tallied out. Lock-up stow should be provided throughout the voyage from the onset of loading to the time of discharge. Shore watchmen and security personnel should be used whenever it is practical and good watch-keeping practice should be the order of the day.

Deep tank use

Many vessels are fitted with 'deep tanks' – employed as ballast tanks or for the carriage of specialized liquid cargoes such as vegetable oils – i.e. coconut oil, bean oil, cotton seed oil, linseed oil, palm oil or mineral oils. Other cargoes include 'tallow' or bulk commodities like grain, molasses or latex.

The specialization of such cargoes often require rigid temperature control of the cargo and to this end most cargo deep tanks are fitted with 'heating coils' which may or may not be blanked off as the circumstances dictate (Figure 3.14).

Note: *Some vessels with a shaft tunnel may be fitted with additional deep tanks aft, in a position either side of the shaft tunnel, but these are not common.*

Preparation of deep tanks

The need for absolute cleanliness with deep tanks is paramount and Cargo Officers are advised that they are virtually always subject to supervision and survey prior to loading example cargoes. Claims for contamination of these cargoes are high and meticulous cleaning of the tank itself and the pipelines employed for loading and discharging must be a matter of course.

Note: *All precautions for the entry into an enclosed space must be taken prior to carrying out maintenance inside 'deep tanks' under a permit to work scheme.*

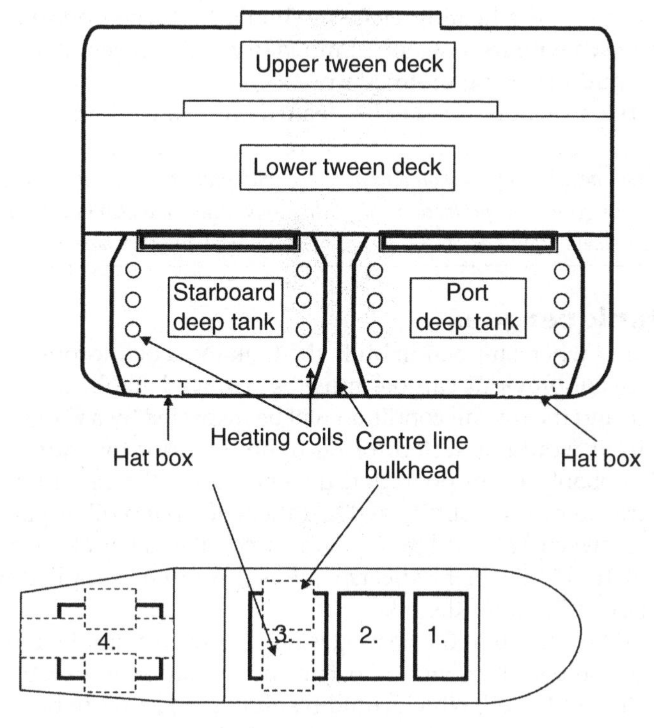

Fig. 3.14 Deep tank storage.

To enable the Classification Surveyor to certify that the tank has watertight integrity and is clean, Chief Officers should, depending on the previous cargo, ensure that:

- After the carriage of a general cargo, the tank is swept down completely and any waste removed.
- In the event of a liquid cargo (assuming of a non-hazardous nature), puddle any residual fluids to the suction and allow the tank to dry.
- If the tank is uncoated (they are often coated in epoxy covering), the bulkhead's decks and deck head should be inspected for rust spots. These should be scraped and wire brushed, and all traces of corrosion removed.
- Heating coils should be rigged and tested. These coils may be 'side coils' or 'bottom coils', or a combination of both.
- Hat boxes should be cleaned out and the suctions should be tested.
- The tank should be filled with clean ballast and the tank lid pressure should be tested (tanks are to be tested to a head of water equal to the maximum to which the tank will be subjected but not less than 2.44 m above the crown of the tank).
- The tank should be emptied to just above the heating coils, a cleansing agent added and the residual water heated by means of the coils. A wash down using a hose and submersible pump then to be carried out.

- After cleaning, the heating element should be turned off and the tank sluiced down with fresh water, pumped dry and allowed to dry, with any residual puddles being mopped up.
- Finally, bilge suctions need to be cleaned and blanked off.

> **Note:** *Personnel so involved should be provided with protective clothing and footwear, together with goggle eye protection. Breathing apparatus may also be a requirement. A risk assessment would be carried out prior to commencing the above task.*

Deep tank cargoes

Vegetable oils – when shipped in bulk, the tank must be thoroughly cleaned and all traces of previous cargoes must be removed. Tank suctions will be blanked off, and the overall condition will be inspected by a Cargo Surveyor. The tank itself would be tested for oil tightness prior to loading. Heating coils will probably be in operation depending on the required shipping temperature. Some oils solidify at 0°C, others like palm oil or palm nut oil, solidify at between 32°C and 39°C, cotton seed oil and kapok seed oil solidify at about 10–13°C. Chief Officers could expect to be supplied with relevant shipping criteria for the oil.

Care must be taken that the heating is not too fierce or applied too quickly as the cargo could scorch. Such an occurrence would be noticeable by some discolouration of the oil, which could result in a cargo claim being filed.

Contamination is avoided by use of shoreside cargo pumps when discharging, while monitoring on passage is conducted by taking ullages and temperatures at least twice per day for oils kept in the liquid state.

Following discharge of the cargo, the tank would probably be steam cleaned and washed with a caustic soda type solution to ensure cleanliness.

Latex – is the 'sap' from rubber trees which rapidly solidifies when exposed to air. It is retained in liquid form by added chemicals, usually ammonia, and shipped in bulk. *Note*: Ammonia attacks brass and copper metalwork and latex tanks should not have such metals as part of their construction.

Prior to loading latex, the tank would be tested and inspected to be thoroughly clean. All steelwork would be coated with hot paraffin wax. The heating coils would be removed as they are not needed for the carriage. Ventilators, air pipes and sounding pipes are all sealed to prevent ammonia loss due to evaporation. Fire extinguishing pipes if fitted should also be plugged. Gas relief valves are fitted to ease any pressure build up inside the tank.

Discharge of the cargo is carried out by shoreside pumps and the tank would then be washed down with water to remove all traces of ammonia. The wax coating is often left in place unless the tank is to be used immediately for another cargo.

Molasses – a syrup obtained from the manufacturing process of sugar. Carried in deep tanks similar to vegetable oils, with heating coils operational to retain the cargo in a liquid state. It is discharged by shoreside pumps and the tanks would be scrubbed and washed down with plenty of water as soon after discharge as is practical. Most contamination claims develop from dirty pipelines. *Note*: Specially designated vessels are employed for the carriage of molasses so the use of deep tanks has diminished with this type of cargo.

Rancidity

The possibility of products turning rancid is always present, especially with fatty oils and fats which contain strong flavours and odours. These elements become developed by being exposed to light, moisture and air, and move towards a condition we know as rancidity. A by-product following excessive exposure and subsequent chemical reaction is the production of fatty acids. These then decompose and form other compounds which are dramatically increased by temperature rise. Such action means that less refined, pure oil is recoverable.

> ***Note:*** *Fats are considered as products which are solid at ordinary temperatures, e.g. 15°C. Fatty oils are those which are liquid at that temperature. The difference between fats and fatty oils is that fatty oils are more chemically reactive than fats.*

Hides – may be shipped in either a wet or dry condition, either in bundles or in casks, or even loose. They are often carried in deep tanks, usually because there is not enough of them to fill a tween deck or lower hold space. Another factor that is against stowage in a tween deck is that wet hides require adequate drainage which would be difficult to achieve in exposed stow. Pickling and/or brine fluid can expect to find its way to the bilges which will necessitate pumping probably twice daily at the beginning of a voyage with hides in the cargo.

> ***Handling precautions*** – *Hides must only be handled with gloves as there is a high risk of contracting anthrax which could prove fatal. Neither should stevedores use hooks in the handling, because of damage to the product. In the case of dry hides these are often brittle and any person being scratched or cut should receive immediate hospital treatment.*

The stowage of hides must be away from dry goods and ironwork. They have a pungent odour and should be stowed well away from other goods that are liable to spoil. They should not be overstowed.

Ballasting and Ballast Management

As cargo is loaded it is general practice for most types of vessels to de-ballast. Some tanks are retained for the purpose of trimming the ship and adjusting the stability conditions, but overall if the ballast was kept on board, the ship could well be seen to be overloaded.

In future it is expected that participating governments to the International Maritime Organization (IMO) convention will have to restrict discharge of ballast water because of the impurities it may contain. To this end a Ballast Management (Record) Book would need to be kept, indicating which tanks are filled/emptied, the position of the Ballast Movement and details of quantities and any treatment, e.g. ultraviolet light which the water may be submitted to.

Stability

When loading/discharging the cargo, due regard must be taken of the ship's condition of stability at every stage and position of the voyage. A reasonable 'GM' must be appropriate throughout the passage and loadline zones must not be infringed. Most modern vessels would engage the flexibility of a 'loadicator' (computer program) for working relevant stability criteria. Associated software of this nature would also provide bending, and shear force stresses incurred and take account of total weights of stores, bunkers, fresh water and ballast contents to provide example conditions.

> ***Note:*** *Free surface moments have a negative effect on the ship's GM, especially when loading or discharging heavy-lift cargoes which may cause the vessel to heel. To this end slack tanks should be avoided if at all possible, when working cargo (see also examples in Chapter 10).*

Loadlines

Ship's Cargo Officers must take care that the vessel is not overloaded beyond the appropriate loadline. Overloading endangers the safety of the vessel and would incur the risk of a heavy fine against the Ship's Master. When loading certain cargoes, especially bulk cargoes like bulk ore and oil the vessel is liable to become hogged or more probably adopt a sagged position. If the vessel is sagging the apparent mean draught will be less than the actual mean draught. This situation does not permit overloading.

The various loadlines (Figure 3.15) are shown and they are assigned to the vessel following a loadline survey by an Assigning Authority, e.g. Lloyds Register.

'S' The summer loadline mark is calculated from the loadline rules and is dependent on many factors including the ship's length, type of vessel and the number of superstructures, the amount of sheer, minimum bow height and so on.

'W' The Winter mark is 1/48th of the summer load draught below 'S'.

'T' The Tropical mark is 1/48th of the summer load draught above 'S'.

'F' The Fresh mark is an equal amount of $\Delta/4T$ millimetres above 'S' where Δ represents the displacement in metric tonnes at the summer load draught and T represents the metric tonnes per centimetre immersion at the above. In any case where the displacement cannot be ascertained, F is the same level as T.

TF The Tropical Fresh mark, relative to 'T' is found in the same manner as that of 'F' relative to 'S'.

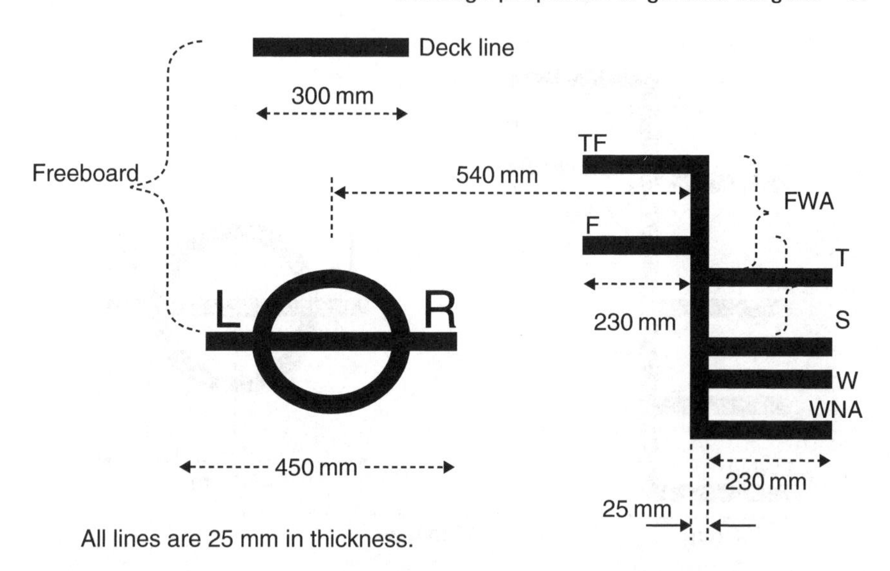

Fig. 3.15 Loadline marks. Should the ship carry a lumber loadline this would be positioned aft of the Plimsoll Mark and identity marks be prefixed with an 'L', e.g. LTF = Lumber Tropical Fresh. LR: Lloyds Register; TF: Tropical Fresh; T: tropical; F: Fresh; S: Summer; W: Winter; WNA: Winter North Atlantic; FWA: Fresh Water Allowance.

WNA The Winter North Atlantic mark is employed by vessels not exceeding 100 m in length when in certain areas of the North Atlantic Ocean, during the winter period. When it is assigned it is positioned 50 mm below the Winter 'W' mark.

Timber loadlines

Certain vessels are assigned Timber Freeboards when they meet certain additional conditions. One of these conditions must be that the vessel must have a forecastle of at least 0.07 extent of the ship's length and of not less than a standard height (1.8 m for a vessel 75 m long or less in length and 2.3 m for a vessel 125 m or more in length, with intermediate heights for intermediate lengths) (Figure 3.16). A poop deck or raised quarter deck is also required if the length of the vessel is less than 100 m. All lines are of 25 mm wide.

LS is derived from the appropriate tables contained in the loadline rules.

LW is one-thirty-sixth (1/36th) of the summer timber load draught below LS.

LT is one-forty-eighth (1/48th) of the summer timber load draught above LS.

LF and Lumber Tropical Fresh (LTF)are both calculated in a similar way to F and TF except that the displacement used in the formula is that of the vessel at her summer timber load draught. If this cannot be ascertained these marks will be one-forty-eighth (1/48th) of LS draught above LS and LT, respectively. LWNA is at the same level as the WNA mark.

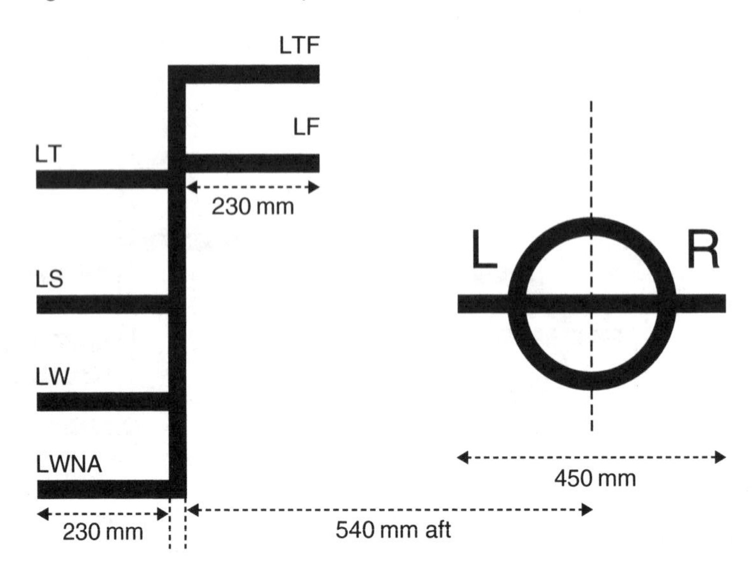

Fig. 3.16 Timber loadlines. The letters denoting the assigning authority LR should be approximately 115 mm in height and 75 mm in width.

Ships with timber loadlines and carrying timber deck cargo in accordance with the M.S. (Loadlines) (Deck Cargo) Regulations 1989 must observe the applicable loadline that she would use if she were not marked with timber loadlines, i.e. Lumber Summer (LS) in the Summer Zone. However, if the timber is not carried in accord with the regulations the ordinary loadlines should be employed.

Note: *The Dock Water Allowance (DWA) would be applied for vessels which are loading in waters other than sea water of 1.025.*

Deadweight scale

Once cargo has been loaded the ship's draughts would normally be ascertained and it would be the Chief Officer's practice to employ the deadweight scale (part of the ship's stability documentation) to ascertain the ship's final displacement. The known figures of fuel, stores and fresh water can then be applied to provide a check against total cargo loaded from the scaled deadweight figure (Figure 3.17).

Offence to overload

Cargo Officers should be aware that it is an offence to overload a vessel beyond her legal marks and attempt to proceed to sea. The owner, or master, will be liable on summary conviction to a fine not exceeding the Statutory maximum of (£5000) or on conviction on indictment, to an unlimited fine. The ship may also be detained until it has been surveyed and marked. The contravention will also carry, in addition to the stated fine, a further £1000 per centimetre of the amount of overload.

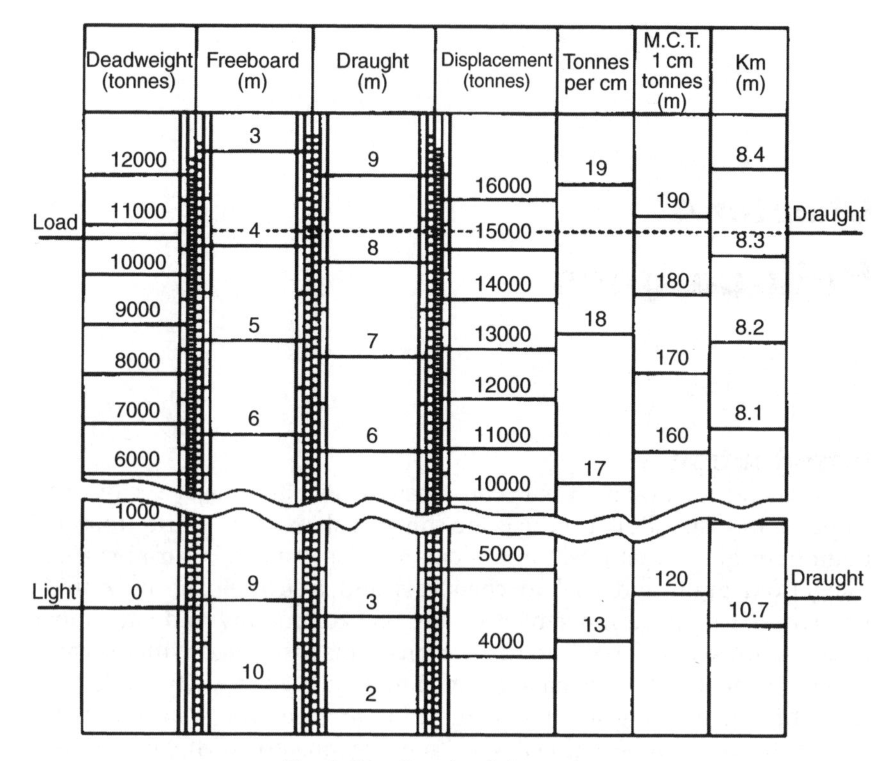

Fig. 3.17 Deadweight scales.

Restrictions to loading

The Loadline Regulations provide various zones around the world's ocean/sea areas. These zones reflect permanent and seasonal areas which are depicted on a chart which accompanies the regulations. There are three permanent zones, namely a *summer zone in each hemisphere* of the globe and a *tropical zone* across the equatorial belt – while the ship is passing through these zones the appropriate loadline would be used.

A ship cannot load deeper than her summer loadline in the summer zone, neither can a vessel load deeper than her tropical mark when in the tropical zone. There are five (5) 'Winter Seasonal Areas', usually found confined by land masses and include: the Black Sea, the Baltic Sea, the Mediterranean, the Sea of Japan and the special 'Winter' area in the North Atlantic, applicable for ships 100 m or less in length.

Cargo Officers will frequently find themselves loading in dock water of less density than sea water and such a situation would warrant use of the DWA formula which would permit a vessel to load beyond her marks, knowing that the vessel will rise to the permitted loadline once entering the sea water of the respective zone or seasonal area.

Chapter 4
Bulk cargoes

Introduction

The demand for raw materials continues to sustain a major sector of the shipping industry. Bulk products are shipped all over the world from their point of origin to that position of demand. The 'bulkers' transport everything from grain and coal to chemicals and iron ore. The bulk trades involve vast tonnage movement of any one commodity and such movement can present its own hazards and problems associated with the cargo.

Designs of ship's holds have evolved to maximize capacity while at the same time generating a safer method of carriage. The Maritime Safety Committee of the International Maritime Organization (IMO) has adopted amendments to Chapter XII Safety of Life at Sea (SOLAS) (Additional Safety Measures for Bulk Carriers) which came into force in July 2004, affecting all bulk carriers regardless of their date of construction. These amendments include the fitting of dry space, water level detectors and alarm monitors, as well as means of draining and pumping, and dry space bilges located forward of the 'collision bulkhead'.

Further recommendations for bulk carriers over 150 m in length to require 'double-hulls' has been agreed (but not yet ratified). Effectively, the double-hull, bulk carrier would seem to be the future for bulk cargoes. How these cargoes are loaded, managed and discharged in the types of vessels involved is as follows.

References for bulk cargoes

International Code for the Safe Carriage of Grain in Bulk.

International Code for the Construction and Equipment of Ships Carrying Dangerous Chemicals in Bulk (IBC Code).

Code of Practice for the Safe Loading and Unloading of Bulk Cargoes (BLU Code).

Code of Safe Practice for Solid Bulk Cargoes (BC Code).

Resolutions of the 1977 SOLAS Conference, regarding the Inspection and Surveys of Bulk Carrier vessels.

MSC/Circ. 908 (June 1999), Appendix C, Uniform Method of Measurement of the Density of Bulk Cargoes.

- MSC/Circ. 646 (June 1994) Recommendations for the Fitting of Hull Stress Monitoring Systems ((also MGN) 108 M).

Definitions and terminology employed with bulk cargoes

Angle of repose – the natural angle between the cone slope and the horizontal plane when bulk cargo is emptied onto this plane in ideal conditions. A value is quoted for specific types of cargoes, results being obtained from use of a 'tilting box'. The angle of repose value is used as a means of registering the likelihood of a cargo shift during the voyage.

An angle of repose of 35° is taken as being the dividing line for bulk cargoes of lesser or greater shifting hazard and cargoes having angles of repose of more or less than the figure are considered separately (Figure 4.1).

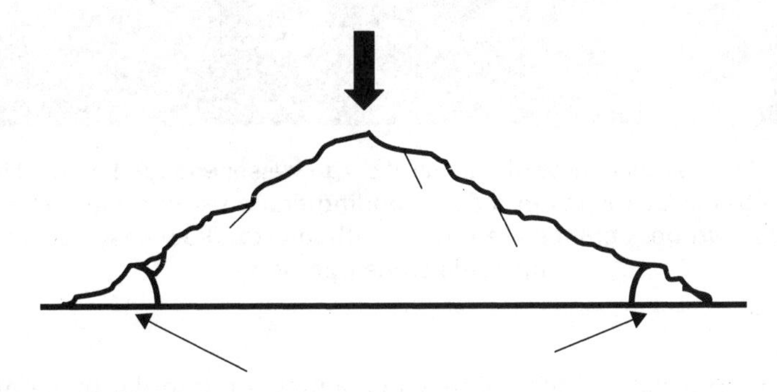

Fig. 4.1 Angle of repose.

Bulk density – is the weight of solids, air and water per unit volume. It includes the moisture of the cargo and the voids whether filled with air or water.

Cargoes which may liquefy – means cargoes which are subject to moisture migration and subsequent liquefaction if shipped with a moisture content in excess of the transportable moisture limit.

Combination carriers (OBO or O/O) – a ship whose design is similar to a conventional bulk carrier but is equipped with pipelines, pumps and inert gas plant so as to enable the carriage of oil cargoes in designated spaces.

Concentrates – these are the materials that have been derived from a natural ore by physical or chemical refinement, or purification processes. They are usually in small granular or powder form (Figure 4.2).

Conveyor system – means the entire system for delivering cargo from the shore stockpile or receiving point to the ship.

Flow moisture point – is that percentage of moisture content, when a flow state develops.

Fig. 4.2 An overhead view of a general cargo vessel engaged in the discharge of concentrates by means of a free-standing crane using a mechanical grab. The ships own deck cranes are turned outboard to allow easy access for the shoreside crane operation.

Flow state – is a state which occurs when a mass of granular material is saturated with liquid to such an extent that it loses its internal shear strength and behaves as if the whole mass was in liquid form.

Incompatible materials – are those materials which may react dangerously when mixed and are subject to recommendations for segregation.

Moisture content – is that percentage proportion of the total mass which is water, ice or other liquid.

Moisture migration – is the movement of moisture contained in the bulk stow, when as a result of settling and consolidation, in conjunction with vibration and the ship's movement, water is progressively displaced. Part or all of the bulk cargo may develop a flow state.

Pour – means the quantity of cargo poured through one hatch opening as one step in the loading plan, i.e. from the time the spout is positioned over a hatch opening until it is moved to another hatch opening.

Transportable moisture limit – the maximum moisture content of a cargo that may liquefy at a level which is considered safe for carriage in ships other than those ships which, because of design features of specialized fittings, may carry cargo with a moisture content over and above this limit.

Trimming – a manual or mechanically achieved adjustment to the surface level of the form/shape of a bulk stow in a cargo space. It may consist of altering the distribution or changing the surface angle to the point, perhaps of levelling some or all of the cargo, following loading.

Code of Safe Working Practice for Bulk Cargoes

(now known as the Bulk Cargo (BC) Code)

The IMO have produced several editions of the code since its conception in 1965. It is meant as a guide and recommendation to governments and shipowners for the carriage of bulk cargoes of various types.

Recommendations are made about the stowage of the cargoes and include suggested maximum weights to be allocated to lower holds as found from the formula:

$$0.9 \times \text{LBD}$$

where L represents the length of the lower hold; B represents the average breadth of the lower hold and D represents the ships summer load draught.

The height of the cargo pile peak should not exceed:

$$1.1 \times \text{D} \times \text{SF (m}^3\text{/tonnes) metres}$$

where SF represents stowage factor.

Legislative, unified requirements (UR) for bulk carriers

Water ingress alarms – are required under SOLAS XII Regulation 12. Such alarms must be fitted to all cargo holds and be audible and visual alarms to the navigation bridge.

Existing bulk carriers are also required to have, in addition to the water level alarms stated above, permanent access for close-up inspection and the use of green sea loads on deck for the design of hatches and deck fittings. Such measures are expected to ensure that a well-maintained single-hull bulk carrier will remain satisfactory for the remainder of its lifetime.

> ***New bulk carriers*** *– The 2004 Design and Equipment meeting of the MSC confirmed the requirement that all bulk carriers would be of 'double-hull' construction (May 2004).* ***Note:*** *The distance between the inner and outer hulls being 1000 mm. (Consultation is still ongoing).*

They will probably also require: harmonized class notation and standard design-loading conditions together with 'double-side shell', water ingress alarms to cargo holds and forward spaces; increased strength and integrity for the foredeck fittings; free fall lifeboats and immersion suits for all crew members (Figures 4.4 and 4.5).

Fig. 4.3 Working bulk cargoes. Mechanical grabs discharging 'scrap metal' from the cargo hold of a small bulk carrier. The single hull construction shows the athwartships, side framing, positioned vertically below the hatch coaming.

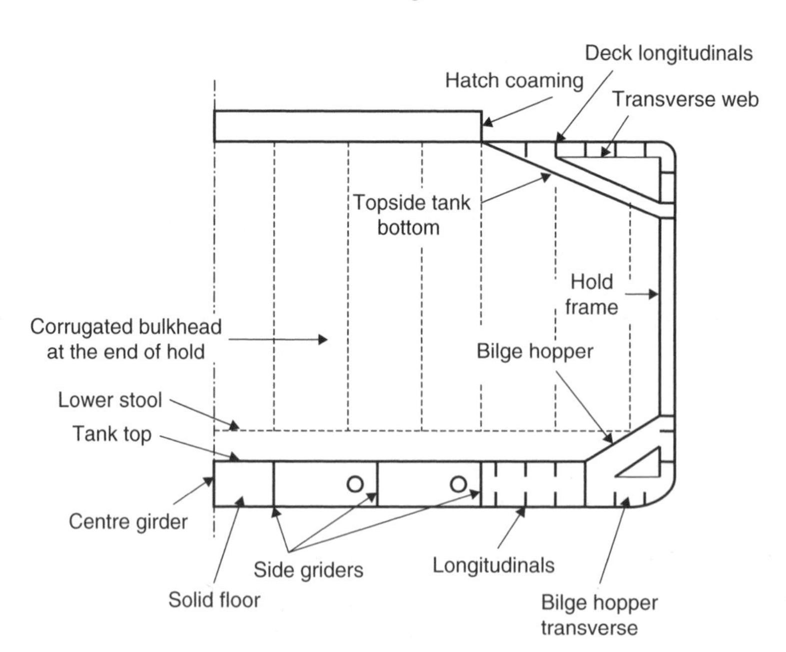

Fig. 4.4 Bulk carrier construction. *Note*: Framing on bulk carriers is designed as a longitudinal system in topside and double bottom tanks and as a transverse system at the cargo hold, side shell position.

Structural changes will also incorporate the permanent means of access for close-up inspection, an amendment to the loadline which will allow the building of stronger and more robust vessels but reduce deadweight capacity by approximate estimates of between 0.5% and 1.5%, depending on size.

Structural standards – as per SOLAS Chapter XII, applying to single-hull side skin bulk carriers, will also apply to new double-hull, bulk carriers.

Additional equipment

At the time of writing, drafted amendments to SOLAS, Chapter XII/II, propose that new bulk carriers over 150 m in length and below shall be fitted with loading instrumentation which provides information on the ship's stability.

Water ingress alarms – are required for vessels with a single cargo hold. The requirement to fit water level detectors in the lowest part of the cargo space is applicable to bulk carriers less than 80 m in length or 100 m in length if built before 1998, to take effect from the first renewal or intermediate survey after July 2004. The alarms will be audible and visual to the navigational bridge and will monitor cargo spaces and other spaces forward of the collision bulkhead. This regulation does not apply to vessels with double sides up to the freeboard deck.

Additional reference

S.I. 1999 M.S. (Additional Safety Measures for Bulk Carriers) Regulations 1999, and MGN 144 (M).

Future builds – double-hulls, bulk carrier construction

The double-hull types have inherent strength that allows flexible-loading patterns, which will increase the capacity for heavy load density cargoes like steel coils. The design dispenses with exposed side frames in the holds and presents a flush side and hold ceiling for cargoes. Such flush features have distinct advantages for hold cleaning and cargo working options with bulk commodities (Figures 4.5 and 4.6).

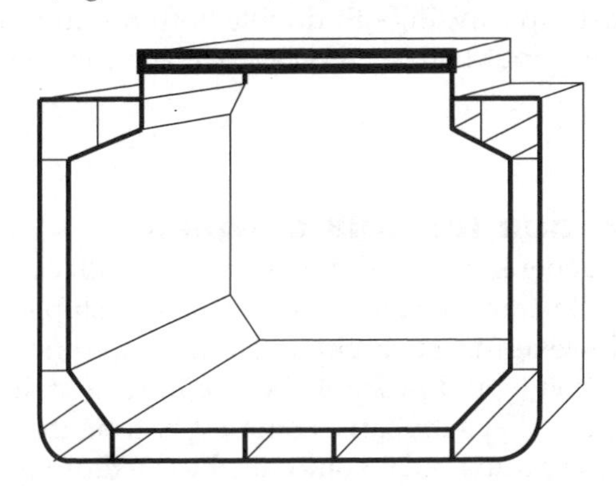

Fig. 4.5 Diamond 53, design – complete double hull in way of cargo holds.

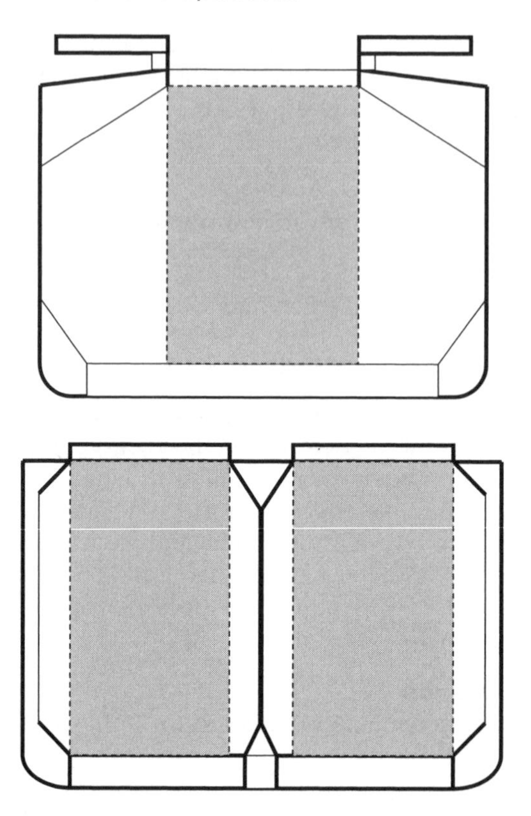

Fig. 4.6 Bulk carrier designs and hatch coverings. (a) Conventional design (twin side moving hatch covers). (b) OJ Libaek's Optimum 2000 (capesize) bulk carrier design (twin hatch covers).

The double design also provides a perceived safer protection against water ingress and is therefore seen as being more environmentally friendly in comparison with the single-hull types. Tank arrangements permit a large water ballast capacity, in both double bottoms and side tanks, eliminating the need to input ballast into cargo spaces, in the event of heavy weather.

Hold preparation for bulk cargoes

Bulk cargoes are generally loaded in designated 'bulk carrier' vessels, but they can be equally transported in general cargo ship's alongside other commodities. However, in such circumstances, specific stowage criteria and hold preparation would probably be a requirement. In virtually every case, except where perhaps the same commodity from the previous voyage is being carried, the cargo holds would need to be thoroughly cleaned and made ready to receive the next cargo.

Designated 'bulk carrier'

1. The holds would be swept down and cleared of any residuals from the previous cargo.
2. All rubbish and waste matter must be removed from the cargo space, before loading of the next cargo can commence.
3. The hold bilge system would need to be inspected and checked to ensure that:
   - the bilge suctions are operational;
   - the bilge bays are clean and smelling sweet (not liable to cause cargo taint).
4. All hold lighting arrangements, together with relevant fittings, would be inspected and seen to be in good order.
5. The space, depending on the nature of the previous cargo and the nature of the next cargo to be carried, would probably require to be washed down with a salt water wash.

> ***Note:*** *Following a wash down, the space would be expected to be allowed to dry out. Special commodities, like foodstuffs, may require the cargo spaces to be surveyed prior to permission being granted to load the ship's cargo.*

Bulk cargoes

Grain

Grain is defined in the IMO Grain Rules as: wheat, maize (corn), rye, oats, barley, rice, pulses or seeds, and whether processed or not, which, when carried in bulk, has a behaviour characteristic similar to grain in that it is liable to shift transversely across a cargo space of a ship, subject to the normal sea-going motion.

> **Applicable to the Grain Rules**
> The following terms mean:
>
> ***Filled*** – *when applied to a cargo space means that the space is filled and trimmed to feed as much grain into the space as possible, when trimming has taken place under the decks and hatch covers, etc.*
>
> ***Partly filled*** – *is taken to mean that level of bulk material which is less than 'filled'. The cargo would always be trimmed level with the ship in an upright condition. Note: A ship may be limited in the number of 'partly filled' spaces that it may be allowed.*

Grain must be carried in accordance with the requirements of the fore-mentioned Grain Rules which consist of three parts, namely 'A', 'B' and 'C'.

Part A Contains 13 rules which refer, among other items, to definitions, trimming, intact stability requirements, longitudinal divisions (shifting boards), securing and the grain-loading information which is to be supplied to the master. This information is to include sufficient data to allow the master to determine the heeling moments due to a grain shift. Thus, there are tables of grain heeling moments for every compartment, which is filled or partly filled, tables of maximum permissible heeling moments, details of scantlings of any temporary fittings, loading instructions in note form and a worked example for the master's guidance.

Part B Considers the effect on the ship's stability of a shift of grain. For the purpose of the rules, it is assumed that in a filled compartment (defined as a compartment in which, after loading and trimming as required by the rules, the bulk grain is at its highest level) the grain can shift into the void space which is always considered to exist at the side of hatchways and other longitudinal members of the structure or shifting boards, where the angle of repose of the grain is greater than 30°.

The average depth of these void spaces is given by the formula:

$$V_d = V_{d1} + 0.75(d - 600) \text{ mm}$$

where V_d represents the average void depth in mm; V_{d1} is the standard void depth found from tables; d is the actual girder depth in mm.

The standard void depth depends on the distance from the hatch end or the hatch side to the boundary of the compartment (Figure 4.7).

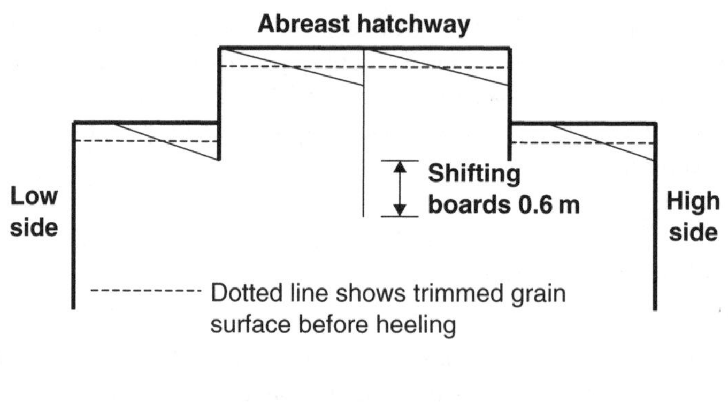

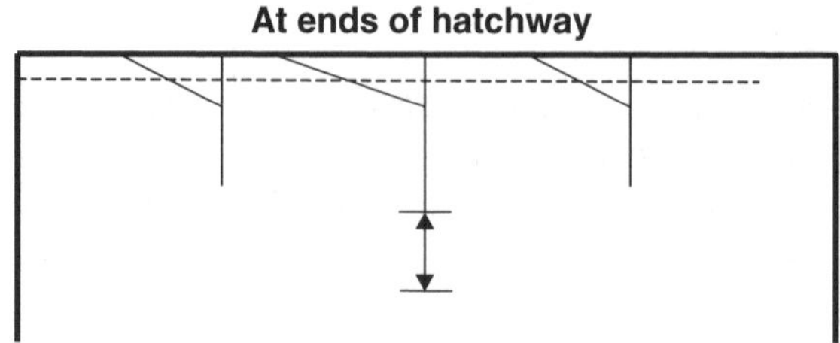

Fig. 4.7 Showing use of shifting boards and assumed formation of voids if heeled 15°.

The assumed transverse heeling moment can now be calculated by taking the product of the length, breadth and half the depth of the void (if it is triangular over the full breadth) and the horizontal distance of the centroid of the void from the centroid of the 'filled' compartment:

The total heeling moment = 1.06 × calculated transverse heeling moment for a full compartment

or

= 1.12 × calculated transverse heeling moment in a partly filled compartment

It will have been noted that the above heeling moments are expressed in m^4 units and so it is also termed a volumetric heeling moment.

The reduction in GZ in the initial position (λ_0) is assumed to be:

$$\frac{\text{Total volumetric heeling moment due to grain shift}}{\text{Stowage factor of the grain} \times \text{displacement}}$$

The reduction in GZ at 40° (λ_{40}) = $0.8\lambda_0$.

Superimposing the above reductions in GZ on the vessel's curve of statical stability will give a 'heeling arm' curve (straight line). The angle at which the two curves cross is the angle of heel due to the shift of grain and this angle must not exceed 12°. Also, the initial metacentric height (GM) (after correction for free surface for liquid in tanks) must not be less than 0.30 m (Figure 4.8).

As can be seen in Figure 4.8, the residual area between the original curve of righting levers and the heeling arms up to 40°, or such smaller angle at which openings in the hull, superstructures or deckhouse cannot be closed watertight immersed (this is called 0_f – the angle of flooding – at which progressive flooding commences) must not be less than 0.075 metre-radians.

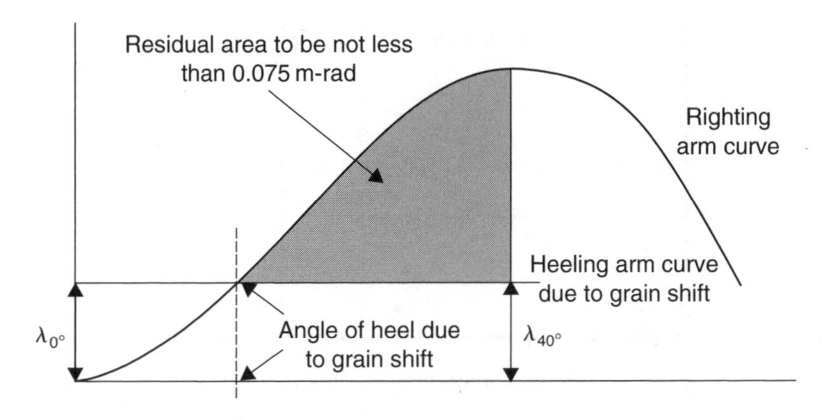

Fig. 4.8 Heeling arm curve.

If the vessel has no Document of Authorization, from the contracting governments, she can still be permitted to load grain if all filled compartments are fitted with centre line divisions extending to the greater of one-eighth maximum breadth of the compartment or 2.4 m. The hatches of filled compartments must be closed with the covers in place. The grain surfaces in partly filled compartments must be trimmed level and secured, and she must have a GM which is to be the greater of 0.3 m or that found from the formula I of the rules.

Part C Concerned with the strength and fitting of shifting boards, shores, stays and the manner in which heeling moments may be reduced by the saucering of grain. The handling of bulk and the securing of hatches of filled compartments and the securing of grain in a partly filled compartment is also detailed.

When shifting boards are fitted in order to reduce the volumetric heeling moment, they are to be of a certain minimum strength with a 15 mm housing on bulkheads and are supported by uprights spaced according to the thickness of the shifting boards (e.g. 50 mm thick, shifting boards would require a maximum spacing of 2.5 m between uprights) (Figure 4.9). The shores will be heeled on the permanent structure of the ship and be as near horizontal as practical but in no case more than 45° to the horizontal. Steel wire rope stays set up horizontally may be fitted in place of wooden shores but the wire must be of a size to support a load in the stay support of 500 kg/m^2.

The shifting boards will extend from deck to deck in a filled tween deck compartment while in a filled hold they should extend to at least 0.6 m below the grain surface after it has been assumed to shift through an angle of 15°.

In a partly filled compartment the shifting boards can be expected to extend from at least one-eighth the maximum breadth of the compartment above the surface of the levelled grain to the same distance below.

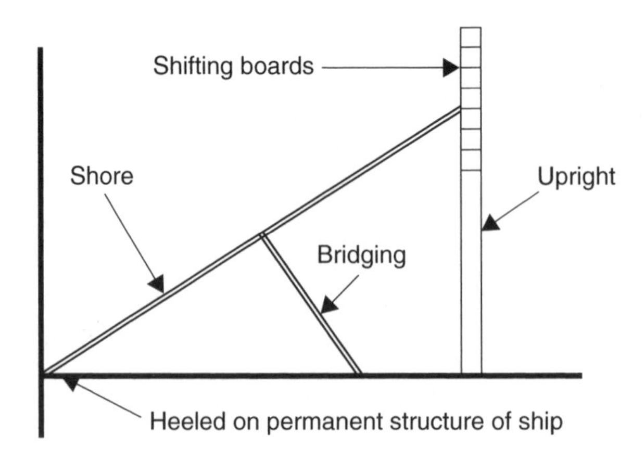

Fig. 4.9 Strength and fitting of shifting boards.

Fig. 4.10 The bulk carrier 'Alpha Afovos' lies port side to the grain silos in Barcelona. The grain elevators seen deployed into the ships hold effecting discharge.

A further method of reducing the heeling moment in a filled compartment is to 'saucer' the bulk in the square of the hatch and to fill the saucer with bagged grain or other suitable cargo laid on separation cloths spread over the bulk grain (Figure 4.12(b)). The depth of the saucer on a vessel over 18.3-moulded breadth will be not less than 1.8 m. Bulk grain may be used to fill a saucer provided that it is 'bundled' which is to say that after lining the saucer with acceptable material, athwartships lashings (75 mm polypropylene or equivalent are placed on the lining material not more than 2.4 m apart and of sufficient length to draw tight over the surface of the grain in the saucer. Dunnage 25 mm in thickness and between 150 and 300 mm wide is laid longitudinally over the lashings). The saucer is now filled with bulk grain and the lashing drawn tight over the top of the bulk in the saucer (Figure 4.11).

In a partly filled compartment, where account is not taken of adverse heeling moments due to grain shift, the surface of the bulk grain is to be trimmed level before being overstowed with bagged grain or other cargo exerting at least the same pressure to a height of not less than one-sixteenth the maximum breadth of the free grain surface or 1.2 m whichever is the greater. The bagged grain, or other suitable cargo, will be stowed on a separation cloth placed overt the bulk grain, or a platform constructed by 25 mm boards laid over wooden bearers not more than 1.2 m apart maybe used instead of separation cloths.

Lashings and bottle screw securings must be regularly inspected and reset taught during the voyage.

Fig. 4.11 Discharge of grain/cereals. The suckers from 'grain elevators' discharge cereals contained in a ships bulk cargo hold from under the rolled back steel hatch covers. Men in the hold actively use the heel ropes to drag the suctions into areas of cereal concentration.

Measures to reduce the volumetric heeling moment of 'filled' and 'partly filled' cargo compartments

- By use of longitudinal divisions – these are required to be grain tight and of an approved scantling.
- By means of a saucer and bundling bulk – a saucer shape is constructed of bulk bundles in the hatch square of a filled compartment. The depth of the saucer being established between 1.3 and 1.4 m depth dependent on the ships beam, below the deck line.
- By overstowing in a partly filled compartments – achieved by trimming the surface level flat and covering with a separation cloth then tightly stowing bagged grain to a depth of one-sixteenth the depth of the free grain stow.

To ensure adequate stability

- The angle of heeling of the vessel which arises from the assumed 'shift of grain' must not exceed 12°.
- When allowing for the assumed shift of grain, the dynamical stability remaining, that is the residual resistance to rolling on the listed side, must be adequate.
- The initial GM, making full allowance for the free surface effect of all partially filled tanks must be maintained at 0.3 m or more.
- The ship is to be upright at the time of proceeding to sea.

Document of Authorization

In order to load grain, a vessel must have a Document of Authorization or an appropriate 'Exemption Certificate'. The authorization means that the vessel has been surveyed and correct grain-loading information has been supplied to the ship for use by the Deck Officer responsible.

Grain awareness

When a grain cargo is loaded, compartments will contain void spaces below the crown of the hatch top. During the voyage the grain will 'settle' and these void spaces would be accentuated. In the event that the grain shifts, it will move into these void spaces to one side or another generating an adverse list to the vessel and directly affecting the stability of the ship by reducing the resistance to roll and adversely affecting the 'Range of Stability'.

Measures to reduce the possibility of the grain shifting include the rigging of longitudinal shifting boards (Figure 4.12(a)), and overstowing the bulk cargo with bagged grain (Figure 4.12(b)).

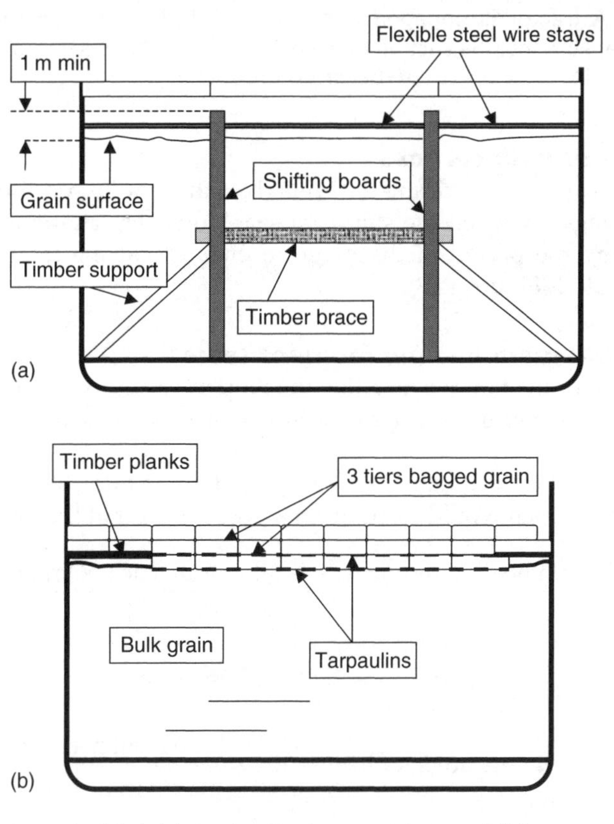

Fig. 4.12 Cargo hold. (a) Longitudinal separation and (b) overstowing bulk cargo with bagged cargo.

Fig. 4.13 A tractor is engaged in the tween deck of a vessel discharging cereals, to ensure that the suction of the grain elevator has access to all residuals of the cargo product.

Trimming of bulk cargoes

Many bulk cargoes are trimmed (levelled) at the loading port to provide a stable stowage for when the ship is at sea. However, trimming also takes place during the period of discharge to ensure that the total volume of cargo is landed (Figure 4.13).

Permissible grain heeling moment tables

The purpose of the tables is to allow the Ship's Master to ascertain whether or not a particular grain stowage condition will achieve the required stability criteria.

The obtained values can then be applied to acquire the approximate angle of heel which would result from a possible shift of the grain cargo.

$$\text{Actual grain heeling moment} = \frac{\text{Total volumetric heeling moment}}{\text{SF}}$$

$$\text{SF (bulk)} = 1.20/1.67\,\text{m}^3/\text{tonne}$$

$$\text{Approximate angle of heel} = 12^\circ \times \frac{\text{Actual heeling moment}}{\text{Permissible heeling moment}}$$

For a ship to be authorized to carry grain, the surveyor will have made calculations for sample cargoes to show that adequate stability for the ship exists. A *grain-loading information listing* should be made in which the surveyor would record all the dimensions of the carriage compartments and then these would be converted into potential heeling moments for when the space is filled or partly filled.

The Deck Officer would be expected to make his own calculations before the intended voyage to take account of the type of grain being carried and its stowage factor. Account must also be taken for the condition of the ship at all stages of the voyage to ensure adequate stability throughout.

Coal – loading, carriage and discharge

Categories of coal

Coal – any coal, including sized grades, small coal, coal duff, coal slurry or anthracite.
Coal duff – coal with an upper size of 7 mm.
Coal slurry – coal with particles generally under 1 mm in size.
Coke – solid residue from the distillation of coal or petroleum.
Small coal – sufficient particle material below 7 mm to exhibit a flow state when saturated with water.

The characteristics of coal

Coal cargoes are liable to spontaneous heating, especially when sufficient oxygen is available to generate combustion. The amount of heating that takes place will depend on the type of coal being carried, and the ability to disperse that heat with effective ventilation methods. Unfortunately, ventilation can work against the safe carriage because of supplying unwanted oxygen, while at the same time dispersing the heat concentrations. It is recommended that surface ventilation only is applied to coal cargoes. This can be applied by raising the hatch tops (weather permitting) to allow surface air and released gases to go to atmosphere and not be allowed to build up inside the cargo compartment.

Freshly mined coal absorbs oxygen which, with extrinsic moisture, forms peroxides. These in turn break down to form carbon monoxide and carbon dioxide. Heat is produced and this exothermic reaction causes further oxidation and further heat. If this heat is not dissipated, ignition will occur, e.g. spontaneous combustion.

Large coal gives a good ventilation path for air flow towards surface ventilation methods, while small coal tends to retain the oxygen content and is more likely to generate spontaneous combustion.

Preparation of the holds should include the overall cleaning of the hold prior to loading, the testing of the bilge suctions and sealing the bilge bays to prevent coal dust clogging bilge bays. Spar ceiling (cargo battens) should also be removed as these would have a tendency to harbour oxygen pockets deep into the heart of the stow. Hold thermometers should be rigged at three different levels, to ensure tight monitoring of the temperatures in the compartments loaded with coal. Critical temperatures in coal vary, but heating will be accelerated in some varieties of coal from as low as 38°C (100°F). Such temperatures would create a need to keep external hull and deck surfaces as cool as possible. In the tropics, it may be appropriate to cover decks to lessen the internal heating in the compartment.

Coal fires

Most coal fires occur at about the tween deck level which is an area that requires more attention to temperature monitoring and to ventilation.

Surface ventilation to holds should be concerned with the removal of gas for the first 5 days of the voyage, thereafter the ventilators to the lower holds should be plugged with an exception for about 6 h every 2 days. Gas from the holds or tween deck regions may find its way into trunk sections, shaft tunnels, chain lockers, peaks and casings unless bulkheads can be maintained in a gas-tight condition.

> ***Note:*** *A strict policy of no naked lights and no smoking should be followed and crew should not be engaged in chipping or painting below decks.*

The majority of coal fires are caused by spontaneous combustion. Poor hatch cleaning prior to loading and a lack of temperature monitoring are often directly linked to the cause. In the event of a coal fire at sea, it should be realized that these are extremely hot fires and if tackled with water would generate copious amounts of steam. Unless this can be vented, the compartments could become pressurized.

If tackled from sea, it is recommended that hatches are battened down and all ventilation to the compartment sealed with the view to starving the fire of oxygen. A Port of Refuge should be sort, where the authorities can be informed to receive the vessel and dig the fire out by grabs while fire-fighters are stood by to tackle the blaze once exposed.

Loading coal

Coal is loaded by either tipping or conveyor belt, bucket system. It is recommended that the first few truck loads are lowered to the holds, this reduces breakage as does a control rate of the chutes. Loading may take place from a single-loading dispenser and, as such, it may become necessary to shift the ship to permit all compartments to be loaded. A loading plan to prevent undue stresses and minimum ship movements would normally be devised. Coal will need to be trimmed as its 'angle of repose' is quite high, especially for large coal.

Small coal like 'mud coal', 'slurry' or 'duff' is liable to shift, but shifting is unlikely in large coal.

Reference should be made to the Code of Safe Working Practice (CSWP) for Bulk Cargoes prior to loading any of the coal types. Information on dry bulk cargoes is given under the heading of 'ores and similar cargoes' and information on wet bulk cargoes is given under the heading of 'ore concentrates'.

Coke

Coke and similar substances such as 'coalite' have had their gas and benzole removed and they do not heat spontaneously. No special precautions are necessary other than to ensure that the coke is cold before loading. If hot coke is loaded this may generate a fire.

The precautions of loading coal

The IMO divides coal into several categories:

Category A – no risk
Category B – flammable gas risk
Category C – spontaneous heating risk
Category D – both risks.

Although precautions are given to each category, the following general precautions are recommended.

1. Gas-tight bulkheads and decks.
2. Spar ceiling (cargo battens) removed.
3. Measures taken to prevent gas accumulating in adjacent compartments.
4. Intrinsically safe electrical equipment inside compartments.
5. Cargo stowed away from high temperature areas and machinery bulkheads.
6. Gas detection equipment on board.
7. Trim cargo level to gain maximum benefits from surface ventilation.
8. Cargo/hatch temperatures monitored at regular intervals.
9. No naked flame or sparking equipment in or around cargo hatches.
10. No welding, or smoking permitted in the area of cargo hatches.
11. Full precautions taken for entry into enclosed compartments carrying coal.
12. Suitable surface ventilation procedures adopted as and when weather permits.

> ***Note:*** *Certain coal cargoes of small particle content are liable to shift if wet, and experience liquefaction hazards. Reference to the IMO Code of Solid Bulk Cargoes should be made and appropriate precautions taken.*

Iron and steel cargoes

Steelwork is carried in various forms, notably as pig iron, steel billets, round bars, pipes, castings, railway iron, 'H' girders, steel coils, scrap metal or iron and steel swarf.

It is without doubt one of the most dangerous of cargoes worked and carried at sea. Recommendations for stowage have been made by various Merchant Shipping Notice (MSNs) and Marine Guidance Notice (MGNs) in the past and yet it is still prone to 'shifting' in a rough sea condition.

Pig iron – if pig iron or billets are taken, they should be levelled and large quantities should not be carried in tween deck spaces. A preferred stow is to level in lower hold spaces and overstow by other suitable cargoes.

If it has to be carried in tween deck spaces the maximum height to which it can be stowed should not exceed 0.22 × the height of the tween deck space.

Pig iron should be trimmed and stowed level in both tween deck and hold spaces in either a side to side or fore and aft stow. If it is not effectively overstowed it should be stowed in robust 'bins', with suitable shifting boards to prevent cargo movement. It is recommended that gloss finished pig iron is always stowed on wood ceiling or dunnage, to reduce steel-to-steel friction.

Round bars and pipes – should preferably be stowed in the lower hold compartments and levelled off. Securing should be in the form of strong cross wires over the top of the stow and secure 'toms' at the sides. Suitable cargoes can overstow this type of steelwork.

Railway iron, 'H' girders, long steel on the round – should be stowed in a fore and aft direction, and packed as solidly as possible. If left exposed and not overstowed, chain lashings should be secured to prevent cargo shift.

Iron and steel swarf – this may heat to dangerous levels while in transit, if the swarf is wet and contaminated with cutting oils. The carriage of 'swarf' requires that surface temperatures of the cargo are monitored at regular intervals during the loading process and whilst on the voyage. If, during loading, the temperature of an area is noted as 48°C (120°F), loading should be temporarily suspended until a distinct fall is observed. In the event that a temperature of 38°C is observed on passage, gentle raking the swarf surface area in the region of the high temperature, to a depth of about 0.3 m should cause the temperature to lower. If a temperature of 65°C is noted, the ship is recommended to make for the nearest port.

Scrap metal – similar problems to other steel cargoes in that it is very heavy. It is generally loaded by elevator/conveyor or grabs and usually discharged by mechanical grabs. When loading, the first few loads are often lowered into the hold to prevent the possibility of excessive damage to ship's structures.

Scrap metal tends to come in all shapes and sizes. As such, where mechanical grabs are engaged, metal pieces frequently become dislodged from the grab when in transit from the hold to the shore, while discharging or loading. Deck Officers should ensure that the working area is cordoned off and personnel on the deck area should wear hard hats and observe cargo operations from a safe distance.

Steel coils – steel coils are stowed on the round and are frequently carried in the cargo holds of 'bulk carriers' (see Chapter 3). The overall stow is secured by steel wire and bottle screw lashings. The sides of the stow are generally chocked tight, against the ship's side, if broken stowage is a feature of the cargo.

Steel coils are classed as a heavy cargo, and would be levelled to no more than two tiers in height. Individually, a coil may weigh up to 10 tonnes, and they are frequently treated as 'heavy lifts'. They are prone to shifting, being stowed on the round, if the vessel encountered rough weather. Passage plans should bear this in mind and chart a 'Port of Refuge' in case such a contingency is required.

Ores – mostly of a low stowage factor, which means that when a full cargo of ore is loaded, there will be a large volume of the hold left unused. A low stowage factor also lends itself to a 'stiff ship', unless some of the cargo can be loaded in the higher regions of the vessel.

Ore should be trimmed if possible and, at the very least, the top of the heap should be knocked off. Modern bulk carrier hold design compensates in some way towards a cargo which is likely to shift. Other vessel designs have been developed as designated ore carriers, and have effective upper ballast compartments to raise the vessel's centre of gravity, when carrying dedicated heavy ore cargoes. In the event that an ore cargo is only a 'part cargo' it should be realized that some ores have a high moisture content which does not always lend to overstowing.

Example ore cargoes: bauxite, chrome, iron, lead and manganese.

Use of mechanical grabs

Many types of bulk cargoes are discharged by means of 'mechanical grabs', of which there are several variations. The handling of grabs is always precarious – especially the larger 5 tonne plus, capacity grabs – because they are not exactly controlled and may incur structural damage to the vessel.

Cargoes tend to be loaded by chute, tipping or pouring, especially the grain type cargoes, ores and coals. However, discharging of ores, bulk solids and the coal cargoes tend to employ grabs for discharge purpose. Bucket grabs coming in various sizes ranging from 2 to 10 tonnes, but the more popular range being in the 4–5 tonne bucket (manual labour for bulk commodity discharge has all but died out) (Figure 4.14).

Cargo Officers are advised that working with heavy grabs requires designated concentration by the crane drivers and even then ship damage is not unusual. A close check on the operation of grabs throughout load/discharge operations is advised and any damage to the ship's structure by contact of the grabs should be reported to the Chief Officer of the vessel. Subsequent damage claims can then be made against the stevedores for relevant repair costs. Bearing in mind that damage to hatch coamings may render the vessel unseaworthy if the damage prevents the closure of hatches and cannot be repaired, before the time of sailing.

Fig. 4.14 Bucket mechanical grab seen in the open position on the quayside. The crane is configured so that the controlling crane driver can open and shut the grab by means of an operational wire to open and close the bucket arrangement.

General information on the loading/discharge of steel cargoes

Steel cargoes in any form are probably one of, if not, the most dangerous cargoes. Steel comes in many forms, from railway lines to ingots, from bulk scrap to bulldozers, etc. It is invariably always heavy and very often difficult to control because of its size. It is a regular cargo for many ships and has been known to cause many problems by way of stability, or adverse effects to the magnetic compass. If steel shifts at sea, due to bad weather, it is unlikely that the crew would have the skills or the facilities to rectify the situation and the vessel would probably need to seek a 'Port of Refuge' with the view to corrective stowage, e.g. steel coils, on the round are particularly notorious for moving in bad weather.

During the loading period, an active Cargo Officer can ensure that correct stowage is achieved and even more important that correct securing is put in place. Relevant numbers of chain lashings and strong timber bearers go well with steel loads, but are often required before the load becomes overstowed by light goods.

Masters should monitor progress during loading periods without being seen to interfere with the Cargo Officer's duties. Specific attention should be given to the use of rigging gangs being employed as and when required. An awareness of the needs of industrial relations without sacrificing the safety requirements can be a delicate balance when a load needs 20 securings and dock labour only wants to secure with 10.

Damage to the vessel when loading or discharging heavy steelwork is not unusual. Heavy lifts by way of bulldozers or locomotives require advance planning and a slow operation. They are awkward to manoeuvre because of either width, length or both. Heavy rig lifting gear operated by ship or shore authorities, even when taking all precautions, very often results in damaged hatch coamings or buckled deck plates. The possibility of damage to the cargo itself is also a likely occurrence.

'H' or 'T' section steel girders are difficult to control because of length and are normally loaded on the diagonal into a hatchway. Slinging is normally by long-leg chains but high winds when loading can cause excessive oscillations of the load, especially with deep sections. Steadying lines of adequate size should be employed before lifting. High winds also pose problems for the lighter steel boilers. These are large but comparatively light, being hollow. Size and shape coupled with strong winds tend to cause slewing on the load in way of the hatch coaming.

Steel in any form will always be shipped and it is in the interests of all concerned to ensure safe handling and stowage. Masters tend to be wary of the stability needs and load in lower holds rather than tween decks depending on circumstances and the needs of other cargoes. However, the need for vigilance when securing remains a high priority towards voyage safety.

Bulk cargo examples

Concentrates – are partially washed or concentrated ores. These cargoes are usually powdery in form and liable to have a high moisture content, and subsequently, under certain conditions, have a tendency to behave almost as a liquid. Special stowage conditions prevail, and sampling must take place to ascertain the transportable moisture limit as provided by the CSWP for Bulk Cargoes.

They are extremely liable to shifting, and care should be taken when loading. Some cargoes may appear to be in a relatively dry condition when loading, but at the same time, contain sufficient moisture to become fluid with the movement and vibrations of the vessel when at sea.

Nitrates – are considered dangerous cargoes. Before stowing, the International Maritime Dangerous Goods (IMDG) Code on dangerous goods should be consulted.

Phosphates – readily absorb water and should be kept dry. A variety of these is Guano which is collected from islands in the Pacific. Phosphates should be kept clear of foodstuffs.

Sulphur – is a highly inflammable cargo and all anti-fire precautions should be taken.

It is also very dusty and highly corrosive. The risk of dust explosions when clearing holds after carriage is of concern. Fires occurring in sulphur cargoes are smothered by use of more sulphur. Personnel should be issued with personnel protection equipment when loading or discharging a sulphur cargo, i.e. masks and goggles.

Nuts – tend to have a high oil content and they are liable to heat and deterioration. They should be kept dry. Precautions should be taken to prevent shifting, as per the grain rules.

Copra – is dried coconut flesh mainly from Malaysia. It is liable to spontaneous heating and is highly inflammable. It is suggested that cargo thermometers are rigged to monitor temperatures in the bulk. Tight anti-fire regulations should prevail around the cargo spaces, to include spark arrester gauze in place on ventilator apertures. The cargo should be kept dry and kept clear of surfaces that are liable to 'sweat'. Matting is recommended to cover the ship's steelwork for this purpose.

Salt – has a high moisture content which is likely to evaporate and dry goods should not be stowed in close proximity. Prior to loading, the spaces should be clean and dry. The steelwork may be whitewashed and separation cloths may be used to keep salt off the ship's structures.

Sugar – vessels have been specifically built for the bulk sugar trade. They are of a similar construction to those of the bulk ore carrier. This is not to say that bulk sugar cannot be carried in any other general type cargo vessel. In any event the compartments should be thoroughly cleaned out and the bilge bays made sugar tight. Bulk sugar must be kept dry. If water is allowed to enter by any means it would solidify the cargo and result in the product being condemned.

Main hazards of loading/shipping/discharging bulk cargoes

Dry shift of cargo – is caused by a low angle of repose and can be avoided by trimming level or the use of shifting boards.

Wet shift of cargo – is caused by liquefaction of the cargo possibly due to moisture migration causing the cargo to act like a liquid, the moisture content of the product probably being below the transportable moisture limit.

Oxidation – the removal of oxygen from the cargo compartment by the type and nature of the cargo, ventilation being required before entry into the compartment.

Flammable/explosive gas/dust – the nature of the cargo has a high risk and may be of a highly inflammable nature, or give off explosive gases. Dusty cargoes also run the risk of a dust explosion of the atmosphere inside the compartment.

Toxic gas or dust – identified toxic effects from products may well require personnel to wear protective clothing and masking/breathing equipment when in proximity of the product.

Corrosive elements (e.g. sulphur) – personnel will require protective clothing, and eye protective wear. High fire risk.

Spontaneous combustion – a self-heating cargo which needs to be monitored by the use of cargo and hatch thermometers throughout the period of the voyage. It should be stowed clear of machinery space bulkheads and provided with recommended ventilation where appropriate (e.g. coal surface ventilation).

Reaction cargoes – products that may react with other cargoes, and as such may require separate stowage compartments.

High density cargoes – may cause structural damage to the vessel and pose stability problems from the position of stow. Could well affect bending and shear force stress effects on the hull.

Infectious cargoes (e.g. Guano) – exposed personnel would require personal protection inclusive of respirators.

Structural damage – through excessive bending and shear forces caused by poor distribution of and/or inadequate trimming of certain cargoes, or sailing with partly filled holds or empty holds.

MARINE GUIDANCE NOTE
MGN 108 (M)

Hull Stress Monitoring Systems

Notice to Shipowners, Ship Operators, Charterers and Managers; Ships' Masters, Ships' Officers, Engineers, Surveyors and Manufacturers of Hull Stress Monitoring Systems.

Summary

- Recommendation for the fitting of hull stress monitoring systems on bulk carriers of 20 000 dwt and above.
- Arrangements for type approval of hull stress monitoring systems.

1. The International Maritime Organization (IMO) recommends the fitting of hull stress monitoring systems to facilitate the safe operation of ships carrying dry cargo in bulk. Use of the system will provide the Masters and Officers of the Ship with real-time information on the motions and global stress the ship experiences while navigating, and during loading and unloading operations. The IMO recommendations are published in the Maritime Safety Committee Circular, MSC/Circ. 646, which is annexed to this Marine Guidance Note.

2. The Maritime and Coastguard Agency (MCA) supports the IMO's recommendations, and invites owners to fit hull stress monitoring systems on bulk carriers of 20 000 dwt and above. Consideration should also be given to fitting such systems to other types of ship.

3. The MCA requests all parties to return information on the reliability of hull stress monitoring systems, their performance relative to the actual and predicted stress levels, their application to other types and sizes of ships and any other relevant experience gained in the use of such systems. Such information should be returned to the Ship Construction Division Section quoting reference MS070/014/0007. This information will be used to inform future deliberations at the IMO, which may include the development of performance standards.

4. The IMO's recommendations call for the hardware and software of the hull stress monitoring system to be approved by the administration. In this respect, the MCA will accept type approval certification of compliance with MSC/Circ. 646, which has been issued by one of the Nominated

Bodies, listed in Table A of the Annex to MSN No. M.1645 "Type Approval of Marine Equipment", who are authorized to examine, test and certify equipment. The terms of M.1645 shall apply. The type approval of hull stress monitoring systems will be included in the next revision of M.1645.

5. Since the adoption of MSC/Circ. 646 in 1994, the design of hull stress monitoring systems has developed and some of the Nominated Bodies are developing standards for such systems. Such development is beneficial and to be encouraged. Consequently, the MCA will accept type approval certification which has been issued by one of the Nominated Bodies in accordance with its published standards or rules, provided that any deviation from MSC/Circ. 646 is recorded on the certificate and notified to the MCA.

6. In designing hull stress monitoring systems, consideration should be given to the IMO Performance Standards for Shipborne Voyage Data Recorders (VDRs), published as Resolution A.861 (2O). Paragraph 5.4.14 of this standard states:

   '5.4.14 Accelerations and hull stresses
   Where a ship is fitted with hull stress and response monitoring equipment, all the data items that have been pre-selected within that equipment should be recorded'

   Owners are invited to ensure that hull stress monitoring equipment is compatible with the VDR fitted and that all monitored data can be transmitted to the VDR.

7. Approval of the hardware and software of the hull stress monitoring system is the first stage but it is essential that the assigning authority for the International Loadline Certificate be consulted regarding the installation of the hull stress monitoring system and the determination of maximum permissible stresses and accelerations. They will also need to be consulted over the frequency of system verification, taking account of the manufacturer's recommendations.

8. Further information on this note may be obtained from:

   Maritime and Coastguard Agency, Spring Place, 105, Commercial Road, SOUTHAMPTON, UK, SO15 lEG. 01703329100 (Tel), 01703 329204 (Fax)

Additional reading and references for bulk cargoes

MGN 144 (M) The Merchant Shipping (Additional Safety Measures for Bulk Carriers), Regulations 1999.

S.I. 1999, No. 336, The Merchant Shipping (Carriage of Cargoes), Regulations 1999.

Chapter 5
Tanker cargoes

Introduction

At the present time modern civilization is largely dependent on oil and its by-products. Vast quantities of liquid products are transported by tankers throughout the world and, as such they have a high profile in the eyes of the general public. However, it should be realized from the outset that not all tankers are in the oil trade. Many transport wine or liquid chemicals, or liquid natural gas (LNG), but generally the tanker vessel is synonymous with the carriage of bulk oil or oil-based products.

Concern for the environment, associated with tanker traffic, has become a number one priority in the anti-pollution campaign and rightly so (Figure 5.1). The marine industry must respect the environment and the well-being of the planet in which we all exist. To this end the Maritime Pollution (MARPOL) convention has gone some way to establishing standards of oil operations around the globe.

Fig. 5.1 A tanker approaches a single buoy mooring (SBM) and prepares to pick up the floating oil pipe with the assistance of local tenders.

The main concern with the demands of a modern society has always been the costs of pollution scaled against societies' needs for oil. Those countries that have it need to go to market to strengthen national economy. While those that are without oil need to import to strengthen their economy. Clearly, an endless circle of world economics. Unfortunately, the tanker accident is not unheard of, e.g. the 'Amoco Cadiz', the 'Torrey Canyon', the 'Exxon Valdez' and the 'Sea Empress' are hard examples to live with.

Our seafarers must be educated – not only to the public outrage that accompanies poor seamanship which generates most modern-day accidents, but also to the ways that prevent such catastrophes happening in the first place. The training of all our seafarers, especially tanker, personnel is an aspect of the marine industry which must take precedence within an industry which continues to drill for oil in the deepest and most remote quarters of the earth.

Definitions for use

(within the understanding of MARPOL) and tanker operations (gas and chemical)

Administration – the Government of the State under whose authority the ship is operating.

Associated piping – the pipeline from the suction point in a cargo tank to the shore connection used for unloading the cargo and includes all the ship's piping, pumps and filters which are in open connection with the cargo unloading line.

Bulk Chemical Code – the Code for the Construction and Equipment of Ships Carrying Dangerous Chemicals in Bulk (ships must have a Certificate of Fitness for the carriage of dangerous chemicals).

Cargo area – that part of a ship which contains cargo spaces, slop tanks and pump rooms, cofferdams, ballast and void spaces adjacent to cargo tanks and also deck areas throughout the length and breadth of the part of the ship over such spaces.

Centre tank – any tank inboard of a longitudinal bulkhead.

Chemical tanker – a ship constructed or adapted primarily to carry a cargo of noxious liquid substances (NLS) in bulk and includes an oil tanker as defined by Annex 1 of MARPOL, when carrying a cargo or part cargo of NLS in bulk (see also Tanker).

Clean ballast – ballast carried in a tank which, since it was last used to carry cargo containing a substance in Category A, B, C or D, has been thoroughly cleaned and the residues resulting therefrom have been discharged and the tank emptied in accord with Annex II, of MARPOL.

Cofferdam – an isolating space between two adjacent steel bulkheads or decks. This space may be a void space or a ballast space.

Combination carrier – a ship designed to carry either oil or solid cargoes in bulk.

Continuous feeding – defined as the process whereby waste is fed into a combustion chamber without human assistance while the incinerator is in normal operating condition with the combustion chamber operative temperature between 850°C and 1200°C.

Critical structural areas – locations which have been identified from calculations to require monitoring or from service history of the subject ship or from similar or sister ships to be sensitive to cracking, buckling or corrosion, which would impair the structural integrity of the ship.

Crude oil – any liquid hydrocarbon mixture occurring naturally in the earth whether or not treated to render it suitable for transportation and includes: (a) crude oil from which certain distillate fractions may have been removed and (b) crude oil to which certain distillate fractions may have been added.

Dedicated ship – a ship built or converted and specifically fitted and certified for the carriage of: (a) one named product and (b) a restricted number of products each in a tank or group of tanks such that each tank or group of tanks is certified for one named product only or compatible products not requiring cargo tank washing for change of cargo.

Domestic trade – a trade solely between ports or terminals within the flag state of which the ship is entitled to fly, without entering into the territorial waters of other states.

Discharge – in relation to harmful substances or effluent containing such substances means any release howsoever caused from a ship and includes any escape, disposal, spilling, leaking, pumping, emitting or emptying.

Emission – any release of substance subject to control by the Annex VI, from ships into the atmosphere or sea.

Flammability limits – the conditions defining the state of fuel oxidant mixture at which application of an adequately strong external ignition source is only just capable of producing flammability in a given test apparatus.

Flammable products – are those identified by an 'F' in column 'F' of the table in Chapter 19 of the International Gas Code for ships carrying liquefied gases in bulk (IGC).

Flash point (of an oil) – this is the lowest temperature at which the oil will give off vapour in quantities that, when mixed with air in certain proportions, are sufficient to create an explosive gas.

Garbage – all kinds of victual, domestic and operational waste, excluding fresh fish and parts thereof, generated during the normal operation of the ship and liable to be disposed of continuously or periodically, except those substances that are defined or listed in other Annexes to the present convention.

Gas carrier – is a cargo ship constructed or adapted and used for the carriage in bulk of any liquefied gas or other products listed in the table of Chapter 19 of the IGC Code.

Good condition – a coating condition with only minor spot rusting.

Harmful substance – any substance that, if introduced into the sea, is liable to create hazards to human health, to harm living resources and marine life, to damage amenities or to interfere with legitimate use of the sea, and includes any substance subject to control by the present convention.

Hold space – is the space enclosed by the ship's structure in which a cargo containment system is situated.

Holding tank – a tank used for the collection and storage of sewage.

IBC Code Certificate – refers to an International Certificate of Fitness for the Carriage of Dangerous Chemicals in Bulk, which certifies compliance with the requirements of the International Bulk Cargo (IBC) Code.

IGC Code – refers to the International Code for the Construction and Equipment of Ships Carrying Liquefied Gases in Bulk.

Ignition point (of an oil) – this is defined by the temperature to which an oil must be raised before its surface layers will ignite and continue to burn.

Incident – any event involving the actual or probable discharge into the sea of harmful substance, or effluents containing such a substance.

Instantaneous rate of discharge of oil content – the rate of discharge of oil in litres per hour at any instant divided by the speed of the ship in knots at the same instant.

International trade – a trade which is not a domestic trade as defined above.

Liquid substances – are those having a vapour pressure not exceeding 2.8 kPa/cm^2 when at a temperature of 37.8°C.

MARVS – is the maximum allowable relief valve setting of a cargo tank.

Miscible – soluble with water in all proportions at wash water temperatures.

NLS Certificate – an international Pollution Prevention Certificate for the Carriage of Noxious Liquid Substances in Bulk, which certifies compliance with Annex II, MARPOL.

Noxious liquid substance – any substance referred to in Appendix II of Annex II of MARPOL. Or, provisionally, assessed under the provisions of Regulation 3(4) as falling into Category A, B, C or D.

NO_x Technical Code – the Technical Code on Control of Emission of Nitrogen Oxides from Marine Diesel Engines, adopted by the Conference, Resolution 2 as may be amended by the Organization.

Oil – petroleum in any form, including crude oil, fuel oil, sludge oil refuse and refined products (other than petrochemicals which are subject to the provisions of Annex II).

Oil fuel unit – is the equipment used for the preparation of oil fuel for delivery to an oil fired boiler, or equipment used for the preparation for delivery of heated oil to an internal combustion engine and includes any oil pressure pumps, filters and heaters with oil at a pressure of not more than 1.8 bar gauge.

Oily mixture – a mixture with any oil content.

Oil tanker – a ship constructed or adapted primarily to carry oil in bulk in its cargo spaces and includes combination carriers and any 'chemical tanker' as defined by Annex II, when it is carrying a cargo or part cargo of oil in bulk (Figure 5.2).

Fig. 5.2 Tanker structure. The 'Jahre Viking' at 564 000 dwt is the largest man-made transport in the world. It is seen manoeuvring with tugs off the Dubai Dry Dock. The size and sophistication of the modern tanker has changed considerably over the decades. World economics have influenced the capacity, while legislation has changed all future construction into the double-hull category. This vessel has recently been converted to a floating oil storage unit to prolong its active life.

Organization – the Inter-Governmental Maritime Consultative Organization. The International Maritime Organization (IMO).

Permissible exposure limit – an exposure limit which is published and enforced by the Occupational Safety and Health Administration (OSHA) as a legal standard. It may be either time weighted average (TWA) exposure limit (8 h) or a 15-min short-term exposure limit (STEL), or a ceiling (C).

Primary barrier – is the inner element designed to contain the cargo when the cargo containment system includes two boundaries.

Product carrier – an oil tanker engaged in the trade of carrying oil, other than crude oil.

Residue – any NLS which remains for disposal.

Residue/water mixture – residue in which water has been added for any purpose (e.g. tank cleaning, ballasting and bilge slops).

Secondary barrier – the liquid resisting outer element of a cargo containment system designated to afford temporary containment of any envisaged leakage of liquid cargo through the primary barrier and to prevent the towering of temperature of the ship's structure to an unsafe level.

Segregated ballast – that ballast water introduced into a tank which is completely separated from the cargo oil and fuel oil system and which is permanently allocated to the carriage of ballast or to the carriage of ballast or cargoes other than oil or noxious substances.

Sewage – (a) drainage and other wastes from any form of toilet, urinals and WC scuppers; (b) drainage from medical premises (dispensary, sick bay, etc.) via wash basins, wash tubs and scuppers located in such premises; (c) drainage from spaces containing living animals and (d) other waste waters when mixed with drainage as listed above.

Ship – a vessel of any type whatsoever operating in the marine environment and includes hydrofoil boats, air cushion vehicles, submersibles, floating craft and fixed or floating platforms (Figure 5.3).

Fig. 5.3 Tanker approaches a floating storage unit (FSU). The pipeline-bearing boom of the FSU is seen in the vertical ready to be deployed once the tanker vessel has moored and connected to the stern of the FSU.

Shipboard incinerator – a shipboard facility designed for the primary purpose of incineration.

Slop tank – a tank specifically designated for the collection of tank drainings, tank washings and other oily mixtures.

Sludge oil – sludge from the fuel or lubricating oil separators waste lubricating oil from main or auxiliary machinery, or waste oil from bilge water separators, oil filtering equipment or drip trays.

Oxides of sulphur (SO_x) emission control area – an area where the adoption of special mandatory measures for SO_x emissions from ships is required to prevent, reduce and control air pollution from SO_x and its attendant adverse impacts on land and sea areas. SO_x emission control areas shall include those listed in Regulation 14 of Annex VI.

Special area – a sea area where, for recognized technical reasons in relation to its oceanographical and ecological condition and to the particular character of its traffic, the adoption of special mandatory methods for the prevention of sea pollution by oil is required. Special areas include Mediterranean Sea, Baltic Sea, Black Sea, Red Sea, Gulf Area, Gulf of Aden, North Sea, English Channel and its approaches, The Wider Caribbean Region and Antarctica.

Substantial corrosion – an extent of corrosion such that the assessment of the corrosion pattern indicates wastage in excess of 75% of the allowable margins, but within acceptable limits.

Suspect areas – are locations showing substantial corrosion and/or are considered by the attending surveyor to be prone to rapid wastage.

Tank – an enclosed space which is formed by the permanent structure of the ship and which is designed for the carriage of liquid in bulk.

Tank cover – the protective structure intended to protect the cargo containment system against damage where it protrudes through the weather deck or to ensure the continuity and integrity of the deck structure.

Tank dome – the upward extension of a position of a cargo tank. In the case of below deck cargo containment system the tank dome protrudes through the weather deck or through a tank covering.

Tanker – an oil tanker as defined by the Regulation 1(4) of Annex 1, or a chemical tanker as defined in Regulation 1(1) of Annex II of the present convention (Figure 5.4).

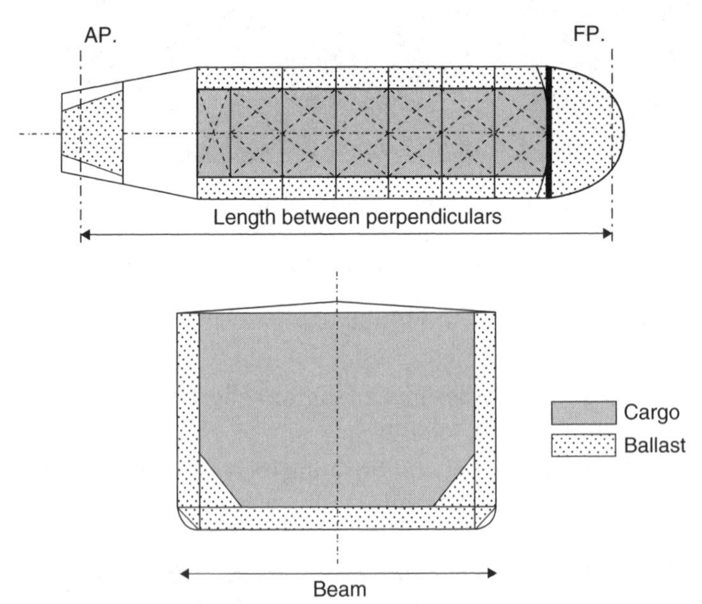

Fig. 5.4 The design of the oil tanker.

Threshold limit value (TLV) – airborne concentrations of substances devised by the American Conference of Government Industrial Hygienists (ACGIH). Representative of conditions under which it is believed that nearly all workers may be exposed day after day with no adverse effects. There are three different types of TLV, TWA, STEL and C. *Note*: TLVs are advisory exposure guidelines, not legal standards and are based on evidence from industrial experience and research studies.

Time weighted average (TWA) – that average time over a given work period (e.g. 8 h working day) of a person's exposure to a chemical or an agent. The average is determined by sampling for the containment throughout the time period and represented by TLV – TWA.

Toxic products – are those identified by a 'T' in column 'F' in the table of Chapter 19 of the IGC Code.

Ullage – that measured distance between the surface of the liquid in a tank and the underside decking of the tank.

Vapour pressure – the equilibrium pressure of the saturated vapour above the liquid expressed in bars absolute, at a specified temperature.

Void space – an enclosed space in the cargo area external to a cargo containment system, other than a hold space, ballast space, fuel oil tank, cargo pump or compressor room, or any space in normal use by personnel.

Volatile liquid – a liquid which is so termed is one which has a tendency to evaporate quickly and has a flash point of less than 60°C.

Wing tank – any tank which is adjacent to the side shell plating.

Equipment regulation requirements

Tankers now require:

- *Cargo tank pressure monitoring* systems required under Safety of Life at Sea (SOLAS) II-2 Regulation 59/IBC Code, Chapter 8.3.3 to be fitted after the first dry docking after 1 July between 1998 and 2002. New build vessels would be similarly equipped.
- *Cargo pump bearing temperature monitoring* systems must be fitted under SOLAS II-2, Regulations 4 and 5.10.1 at the next dry docking after 1 July 2002.
- *Cargo pump gas detection/bilge alarm* systems are now required under SOLAS II-2, Regulations 4 and 5.10.3/5.10.4 at the next dry docking after 1 July 2002.
- *High level and overfill alarm* system is now required under United States Coast Guard (USCG) Regulation 39.
- *Emergency escape breathing devices (EEBDs)* are now required under SOLAS II-2, Regulation 13.3.4 by the first survey after 1 July 2002.

Double-hull tanker construction

Tankers were generally constructed with either centre tanks and wing tanks dividing the vessel into three athwartships sections, by two longitudinal bulkheads, individual tanks being segregated by transverse bulkheads. Modern construction, which integrates the double hull, has meant that construction designs have changed and twin tanks are now positioned to either side of a centre line bulkhead (Figure 5.5).

Fig. 5.5 Athwartships cross-section of the modern double-hull tanker seen at a late construction stage prior to assembly.

The maximum length of an oil tank is 20% L (L represents the ship's length) and there is at least one wash bulkhead if the length of the tank exceeds 10% L or 15 m. It should be appreciated that in a large tanker of 300 m length and 30 m beam and equivalent depth, each tank would have a capacity for over 20 000 tonnes of oil.

Tanks are usually numbered from forward to aft with pump rooms usually situated aft so that power can be easily linked direct from the engine room. Pipeline systems providing flexibility in loading/discharging interconnecting the tanks to the pumping arrangement.

Tanker pipelines

There are three basic types of pipeline systems:

1. Direct system
2. Ring main system
3. Free flow system.

Each system has their uses and is designed to fulfil a need in a particular type of vessel.

The direct system

This is the simplest type of pipeline system which uses fewer valves than the others. It takes oil directly from the tank to the pump and so reduces friction. This has an affect of increasing the rate of discharge, at the same time improving the tank suction. It is cheaper to install and maintain than the ring main system because there is less pipeline length and with fewer valves less likelihood of malfunction. However, the layout is not as versatile as a ring main system and problems in the event of faulty valves or leaking pipelines could prove more difficult to circumvent. Also, the washing is more difficult since there is no circular system and the washings must be flushed into the tanks Figure 5.6.

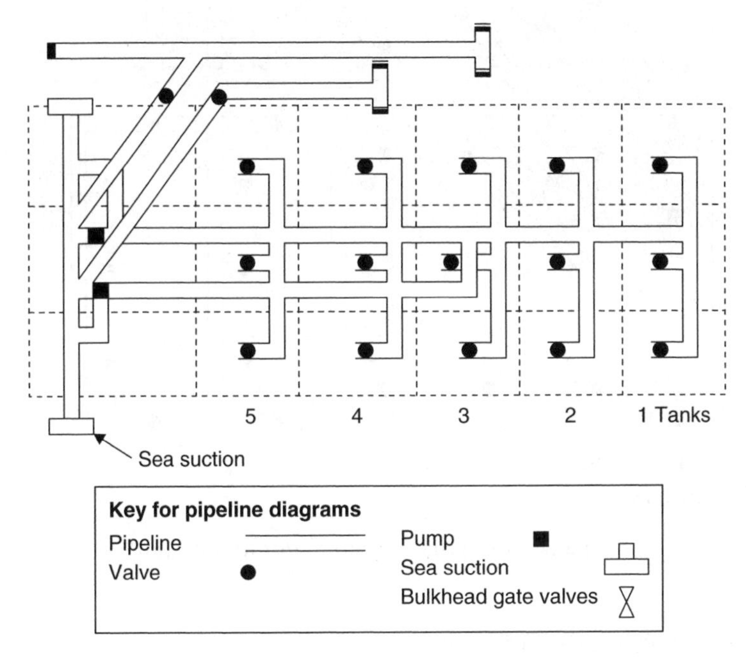

Fig. 5.6 Direct line system. Used mainly on crude and black oil tankers where separation of oil grades is not so important.

The advantages are that:

1. it is easy to operate and less training of personnel is required
2. as there are fewer valves it takes less time to set up the valve system before commencing a cargo operation
3. contamination is unlikely, as it is easy to isolate each section.

The disadvantages are that:

1. it is a very inflexible system which makes it difficult to plan for a multi-port discharge
2. block stowage has to be used which makes it difficult to control 'trim'
3. carrying more than three parcels concurrently can be difficult.

The ring main system

This is basically a ring from the pump room around the ship, with crossover lines at each set of tanks. There are various designs usually involving more than one ring. It is extensively employed on 'product tankers' where the system allows many grades of cargo to be carried without contamination. This is a highly versatile system which permits several different combinations of pump and line for any particular tank (Figure 5.7).

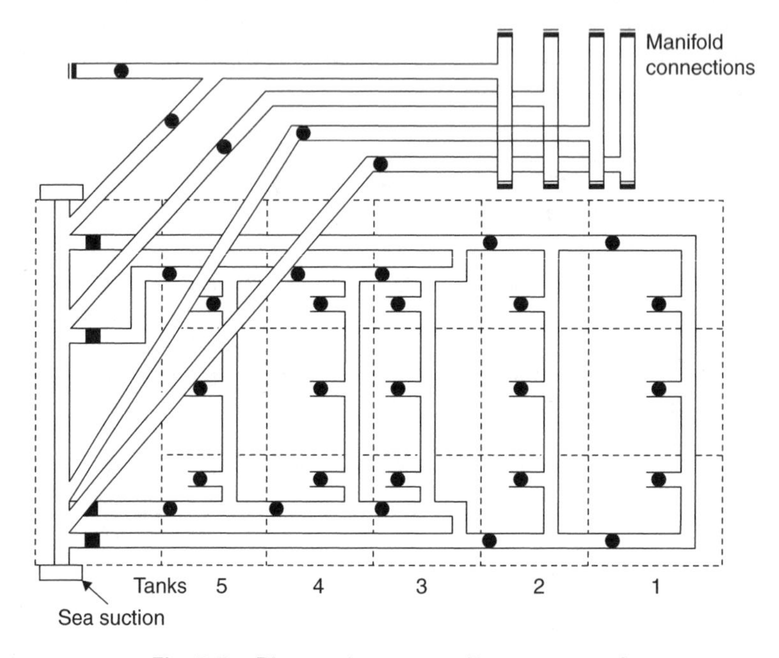

Fig. 5.7 Ring main system. Pump room aft.

The advantages of the system are that:

1. cargoes can be more easily split into smaller units and placed in various parts of the ship
2. line washing is more complete
3. a greater number of different parcels of cargo can be carried
4. trim and stress can be more easily controlled.

The disadvantages are that:

1. because of the more complicated pipeline and valve layout, better training in cargo separation is required
2. contamination is far more likely if valves are incorrectly set
3. fairly low pumping rates are achieved
4. costs of installation and maintenance are higher because of more pipeline and an increased number of valves.

The free flow system

The 'free flow system' employs sluice valves in the tank bulkheads rather than pipelines. With a stern trim this system can discharge all the cargo from the aftermost tank via direct lines to the pump room. The result is that a very high speed of discharge can be achieved and as such is suitable for large crude carriers with a single grade cargo. Tank drainage is also very efficient since the bulkhead valves allow the oil to flow aft easily. There are fewer tanks with this system and it has increased numbers of sluice valves the farther aft you go. The increased number of sluices is a feature to handle the increased volume being allowed to pass from one tank to another (Figure 5.8).

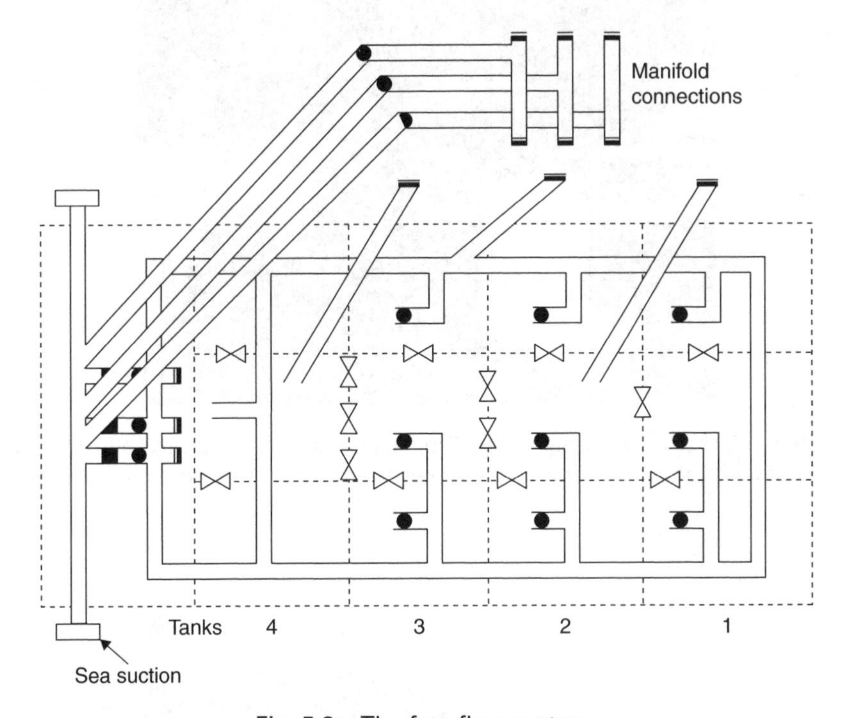

Fig. 5.8 The free flow system.

The main advantage is that a very high rate of discharge is possible with few pipelines and limited losses to friction. The main disadvantage is that overflows are possible if the cargo levels in all tanks are not carefully monitored (Figures 5.9, 5.10 and 5.11).

Measurement of liquid cargoes

The volume of oil in a tank is ascertained by measuring the distance from a fixed point on the deck to the surface of the oil. The distance is known as the 'ullage' and is usually measured by means of a plastic tape. A set of tables is supplied to every ship, which indicate for each cargo compartment, the volume of liquid corresponding to a range of ullage measurements. The

Fig. 5.9 Tanker deck arrangement. Typical example of the pipeline (fore deck) arrangement of a medium size oil tanker seen in the sea-going environment.

Fig. 5.10 Manifold and pipeline connections. Upper deck of an oil tanker showing the manifold, Samson Posts positioned to Port and Starboard, fitted with 5-tonne safe working load (SWL) derricks for lifting hoses to manifold connections.

Fig. 5.11 Manifold and pipeline connections. Typical 14-inch oil pipe connection to the manifold.

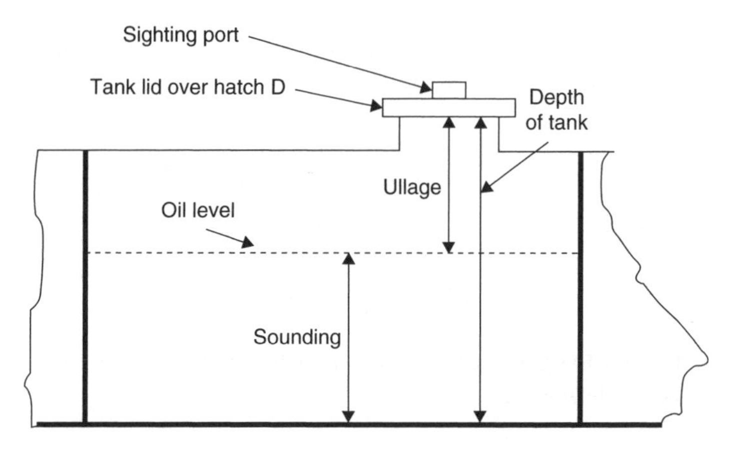

Fig. 5.12 Measurement of liquid cargoes.

ullage opening is usually set as near as possible to the centre of the tank so that for a fixed volume of oil, the ullage is not appreciably affected by conditions of trim and list. If a favourable siting is not possible then the effects of list and trim should be allowed for (Figure 5.12).

The important measure of oil is weight and this must be calculated from the volume of oil in each tank. Weight in tonnes is quickly found by multiplying the volume of oil in cubic metres by the relative density (RD) of the oil. This density is a fraction and may be taken out of petroleum tables when the RD of the oil is known.

Example

To find the weight of 125 m^3 of oil at a RD of 0.98.

$$\begin{aligned}\text{Density of oil} &= 0.98\,t/m^3 \\ \text{Weight of oil} &= \text{volume} \times \text{density} \\ &= 125\,m^3 \times 0.98 \text{ tonnes} \\ &= 122.5 \text{ tonnes}\end{aligned}$$

Oil expands when heated and its RD, therefore, decreases with a rise in temperature. In order that the weight may be calculated accurately, it is important that when ullages are taken the RD of the oil should also be known. This may be measured directly, by means of a hydrometer.

The RD of a particular oil may be calculated if the temperature of the oil is taken. The change of RD due to a change of 1°C in temperature is known as the RD coefficient. This lies between 0.0003 and 0.0005 for most grades of oil and may be used to calculate the RD of an oil at any measured temperature if the RD at some standard temperature is known.

Examples

A certain oil has an RD of 0.75 at 16°C. Its expansion coefficient is 0.00027/°C. Calculate its RD at 26°C.

$$\begin{aligned}\text{Temperature difference} &= 26°C - 16°C = 10°C \\ \text{Change in RD} &= 10 \times 0.00027 \\ &= 0.0027 \\ \text{RD at 16°C} &= 0.75 \\ \text{RD at 26°C} &= 0.7473\end{aligned}$$

An oil has an RD 0.75 at 60°F. Its expansion coefficient is 0.00048/°F. Calculate its RD at 80°F.

$$\begin{aligned}\text{Temperature difference} &= 80°F - 60°F = 20°F \\ \text{Change in RD} &= 20 \times 0.00048 \\ &= 0.0096 \\ \text{RD at 60°F} &= 0.75 \\ \text{RD at 80°F} &= 0.7404\end{aligned}$$

Tank measurement and ullaging

Use of the Whessoe Tank Gauge

The function of the gauge is to register the ullage of the tank at any given time, in particular when the liquid level in the tank is changing during the loading and discharge periods. The gauge is designed to record the readings not only at the top deck level of the tank but also remotely at a central cargo control room. A transmitter is fitted on the head of the gauge for just this purpose.

The unit is totally enclosed and various models manufactured are suitable for use aboard not only oil tankers, but chemical and gas carriers as well (Figure 5.13).

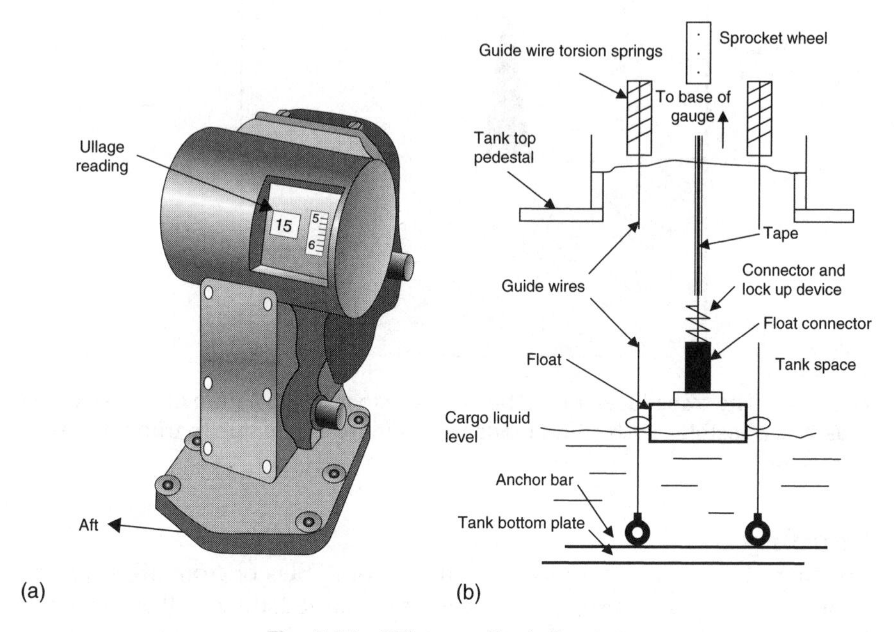

Fig. 5.13 Whessoe Tank Gauge.

Inside the gauge housing is a calibrated ullage tape, perforated to pass over a sprocket wheel and guided to a spring-loaded tape-drum. The tape extends into the tank and is secured to a float of critical weight. As the liquid rises or falls, the tape is drawn into, or extracted out from, the drum at the gauge head. The tape-drum, being spring loaded, provides a constant tension on the tape, regardless of the amount of tape paid out. A counter window for display is fitted into the gauge head, which allows the ullage to be read on site at the top of the tank.

Tank measurement – radar system

This is a totally enclosed measuring system which can only be employed if the tank is fully inerted. Systems are generally fitted with oxygen sensor and temperature sensor switches, so if the atmosphere in the tank is hot or flammable the radar will not function.

The main unit of the system is fitted on the deck with an inserted cable tube into the tank holding a transducer. Cable then carries the signal to a control unit in the cargo control room where the signal is converted to give a digital read-out for each tank monitored (Figure 5.14).

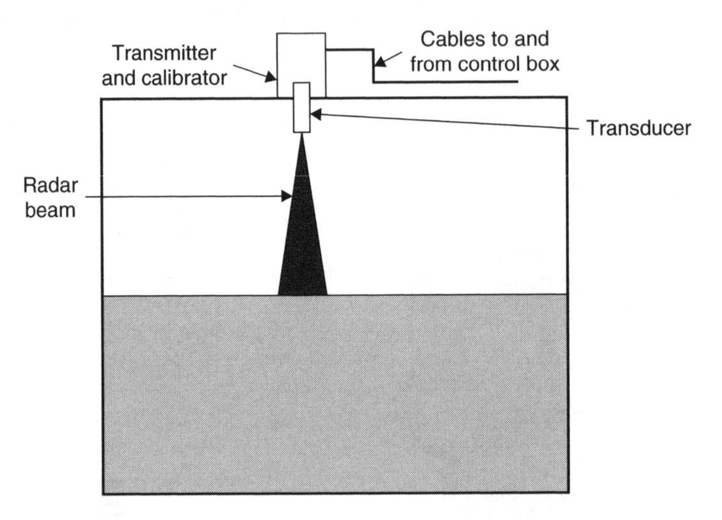

Fig. 5.14 The transducer would be fitted as close to the centre of the tank area as was possible. Such siting tends to eliminate errors due to trim and list.

Loading

Loading of tankers takes place at jetties, from FSUs or from SBM. Where booms carrying oil-bearing pipes are to be connected, these will be insulated to prevent stray currents flowing, as from corrosion prevention systems employed on both ships and jetties. The flow of current in itself should not be a problem, but it may give rise to a spark when making or breaking connections to the manifold. For this reason, these sections are tested regularly for efficient insulation. Lines are often bonded to reduce static electricity effects which could also give rise to an unwanted source of ignition from the fast pumping of liquids (Figure 5.15).

These points are highlighted to illustrate that a high degree of awareness is required in all tanker operations whether loading, discharging or gas freeing. Fire precautions are paramount because the risk of fire aboard the tanker is a real hazard and stringent fire precautions must be adopted throughout cargo operations of every kind.

Fig. 5.15 Moorings and floating oil-bearing pipeline seen extending from the FSU 'Zapro Producer'.

Loading procedural checklist

Company policy on loading procedures vary and Cargo Officers should adhere to the company procedures and take additional reference from the International Safety Guide for Oil Tankers and Terminals (ISGOTT):

1. Complete and sign the ship/shore checklist
2. Establish an agreed communication network
3. Agree the loading plan by both parties and confirm in writing
4. Loading and topping off rates agreed
5. Emergency stop procedures and signals agreed
6. All effected tanks, lines, hoses inspected prior to commencing operations
7. Overboard valves sealed
8. All tanks and lines fully inerted
9. Inert gas (IG) system shut down
10. Pump room isolated and shut down
11. Ships lines set for loading
12. Off side manifolds shut and blanked off
13. All fire fighting and Ships Oil Pollution Emergency Plan (SOPEP) equipment in place
14. Notice of readiness accepted
15. First set of tanks and manifold valves open
16. Commence loading at a slow rate
17. Check and monitor the first tanks to ensure cargo is being received
18. Carry out line sample
19. Check all around the vessel and overside for leaks

20. Increase loading rate to full
21. Check ullages at half-hourly intervals and monitor flow rate to confirm with shoreside figures
22. Check valves operate into next set of tanks prior to change over
23. Reduce loading rate when topping off final tank
24. Order stop in ample time to achieve the planned ullage/line draining
25. When the cargo flow has completely stopped close all valves
26. After settling time, take ullages, temperatures and samples
27. Ensure all log book entries are completed
28. Cause an entry to be made into the Oil Record Book.

> ***Note:*** *The loading plan devised by Chief Officers and Shoreside Authorities would take account of the ship's stability and the possibility of stresses being incurred during all stages of the loading procedure.*

Load on top

When a crude oil tanker completes discharge, a large quantity of oil (upto 2000 tonnes) may be left adhering to the bulkheads. The 'load on top' principle is a method designed to gather all this oil and deposit it into a slop tank. Tank cleaning would be carried out in the normal way drawing in sea water from either a ballast tank or directly from the sea suction.

On completion of tank cleaning the slop tank will contain all the tank washings, made up of a mixture of oil and water (probably in the ratio of three parts water to one part oil). This mixture will contain small particles of oil held in suspension in the water and water droplets will be suspended in the oil. For this reason the slop tank must be allowed to 'settle' for up to about 2 or 3 days. After this period of time the oil can be expected to be floating on top of the water content.

Once settling out is completed the interface between the oil and the water levels must be determined (usually carried out by an interface instrument). Once the level of water is known, it is now possible to estimate the amount of water which can be discharged. The pumps and pipelines would be cleaned of oil particles and the water in the tank can be pumped out very carefully as the interface approaches the bottom. The main cargo pump is stopped when the water depth is at about 15–25 cm.

Alternative methods could be to pump the whole of the slop tank contents through an oily water separator or the tank can be de-canted from one tank to another.

On arrival at the loading port the new hot oil can be loaded on top of the remaining slops, which would have been quantified prior to commencing loading of the new cargo. During the loaded passage the old and new oils combine and any further water content sinks to the bottom of the tank.

On arrival at the discharge port, water dips are taken and the water quantity calculated. This is then usually pumped direct to a shoreside slop tank. Once pure oil is drawn this can be diverted to main shoreside oil tanks.

The main purpose of 'load on top' is to reduce the possibility of oil pollution while the vessel is at sea while at the same time as carrying out a full tank-cleaning programme.

> ***Note:*** *Tank washings containing any persistent oil must not be disposed of into the sea inside territorial waters or 'special areas'.*

Loading capacity

The amount of cargo a tanker can lift will depend upon the vessel's deadweight when the vessel is floating at her designated loadline. The amount of bunkers, fresh water and stores would be deducted to give the total weight of cargo on board. The order of loading tanks is of high priority in order to avoid excessive stresses occurring. Visible damage might not be an immediate result of a poor loading sequence but subsequent damage may be caused later, when in a seaway, which could be attributable to excessive stresses during loading periods.

Nowadays vessels are equipped with designated 'loadicators' or computer software programs to establish effective loading plans and show shear forces and bending moments throughout the ships length. Such aids are beneficial to Ship's Officers in illustrating immediate problems and permitting ample time to effect corrective action.

Although a high rate of loading is usually desirable, this in itself generates a need for tight ship keeping. Moorings will need to be tended regularly and an efficient gangway watch should be maintained. Communications throughout the loading period should be effective and continuous with shoreside authorities, with adequate notice being given to the pumping station prior to 'topping off'.

Care during transit

It would be normal practice that, through the period of the voyage, regular checks are made on the tank ullage values and the temperatures of all tanks. Empty tanks and cofferdams, together with pump rooms, should be sounded daily to ensure no leakage is apparent. Generally, oil is loaded at a higher temperature than that which will be experienced at sea, as such it would be expected that the oil will cool and the ullage will increase for the first part of the voyage.

Viscous oils like fuel oil or heavy lubricating oil would normally be expected to be heated for several days before arrival at the port of discharge. Heating will decrease the viscosity and a higher rate of discharge can be anticipated. Overheating should be avoided as this could affect the character of the product and may strain the structure of the vessel.

Tanks are vented by exhaust ventilators above deck level via masts and Samson Posts. Volatile cargoes such as 'gasoline' are vented via pressure relief valves which only operate when the tank pressure difference to atmosphere exceeds 0.14 kg/cm^2. This prevents an excessive loss of cargo due to evaporation. Evaporation of cargo can also be reduced in hot weather by spraying the upper decks cool with water.

Discharging

Flexible hoses are connected to the ship's manifold, as at the loading port, and the ship to shore checklist would be completed. Good communications between the ship and the shore authority is essential. All overboard discharges should be checked and if all valves are correct, discharge would be commenced at an initial slow rate. This slow rate is commenced to ensure that if a sudden rise in back pressure is experienced in the line, the discharge can be stopped quickly. Such an experience would probably indicate that the receiving lines ashore are not clear.

Back pressure should be continually monitored during discharge operations and the ship, using ship's pumps, should be ready to stop pumping at short notice from a signal from the terminal. The waterline around the ship should also be kept under regular surveillance in the event of leakage occurring.

As with loading operations, the deck scuppers should all be sealed and SOPEP recommendations followed. All fire-fighting equipment should be kept readily available throughout the operation.

Ballasting

In order that no oil is allowed to escape into the sea when engaging in ballast operations, the pumps should be started before the sea valves are opened. If it is intended to ballast by gravity it is still preferable to pump for the first 10 min or so to ensure that no oil leaks out.

Care should also be taken when topping up ballast tanks since any water overflow could be contaminated with oil. Any gas forced out of tanks during ballast operations constitutes a fire risk as equally dangerous as when loading.

All ballast operations should be recorded in the Ballast Management Record Book and any transfers of oil content should be recorded in the Oil Record Book. Log books should take account of all tank operations regarding loading, discharging, ballasting or cleaning.

Tank-cleaning methods

There are generally three methods of cleaning tanks:

1. Bottom flushing with water, petroleum product or chemical solvents
2. Water washing (hot or cold) employing tank-washing machinery
3. Crude oil washing (COW).

Bottom washing

Bottom flushing is usually carried out to rid the tank bottoms of previous cargo prior to loading a different, but compatible grade of cargo.

It can be effective when carrying refined products in small quantities. Bottom washing with acceptable solvents is sometimes conducted, especially where a tanker is to take say paraffin (kerosene) products after carrying leaded gasolines. It should be realized that bottom washing will not remove heavy wax sediments from the bottom of tanks and is used purely as a means of removing the traces of previous cargo.

Portable or fixed washing machines

Using a high-pressure pump and heater, sea water, via a tank-cleaning deck line, is applied to wash the tank thoroughly. The dirty slop water is then stripped back to the slop tank where it is heated to separate the oil from the water.

This is considered an essential method when changing trades from carrying crude to the white oil cargoes, or when the tank is required for clean ballast, or if it is to be gas freed.

COW

A procedure that is conducted during the discharge and which has positive advantages over water-washing methods. New crude oil carriers over 20 000-dwt tonnes must now be fitted and use a COW facility. The method employs a high-pressure jet of crude oil from fixed tank-cleaning equipment. The jet is directed at the structure of the tank and ensures that no slops remain onboard after discharge, every last drop of cargo-going ashore. The advantages are that tank cleaning at sea is avoided, with less likelihood of accidental pollution; less tank corrosion is experienced than from water washing; increased carrying capacity is available for the next cargo; full tank drainage is achieved; and time saved gas freeing for dry dock periods.

Some disadvantages of the system include crew workload, which is increased at the port of discharge; discharge time is increased; it has a high installation cost and maintenance costs are increased, while crew need special training with operational aspects.

Aspects of COW

The operational principle of the COW system is to use dry crude from a full tank to wash the tanks being discharged. Crude containing water droplets from the bottom of a tank should not be used for washing purposes as this may introduce water droplets that have become electrostatically charged and produce an unnecessary source of ignition in the tank atmosphere.

To this end any tank designated for use as COW should be first de-bottomed into the slop tank or bled ashore with the discharge pump.

One of the main cargo pumps is used to supply the COW line with pressurized crude for washing operations. The line, along the deck, will carry branch lines to all of the fixed machines. Large and very large crude carrier (VLCC) vessels may have up to six (6) machines per tank.

Safety in operation Tanks must be fully inerted prior to commencing washing operations and the heater in the tank-washing system must be isolated by blanks. The line would need to be pressurized and tested for leaks prior to commencing washing.

- *Operation – Stage One*: The limits to cover the top of the cycle would need to be adjusted to be pointing upwards. Where portable drive units are employed these would have to be initially fitted and limits set accordingly (Figure 5.16(a)).
- *Operation – Stage Two*: The second stage starts when one-third of the tank is discharged and the washing jet will only be allowed to travel down to a point where the jet strikes the bulkhead just above the level of the oil in the tank. At this stage the machine completes 1½ cycles and must therefore be adjusted, up again, before the start of the next stage (Figure 5.16(b)).

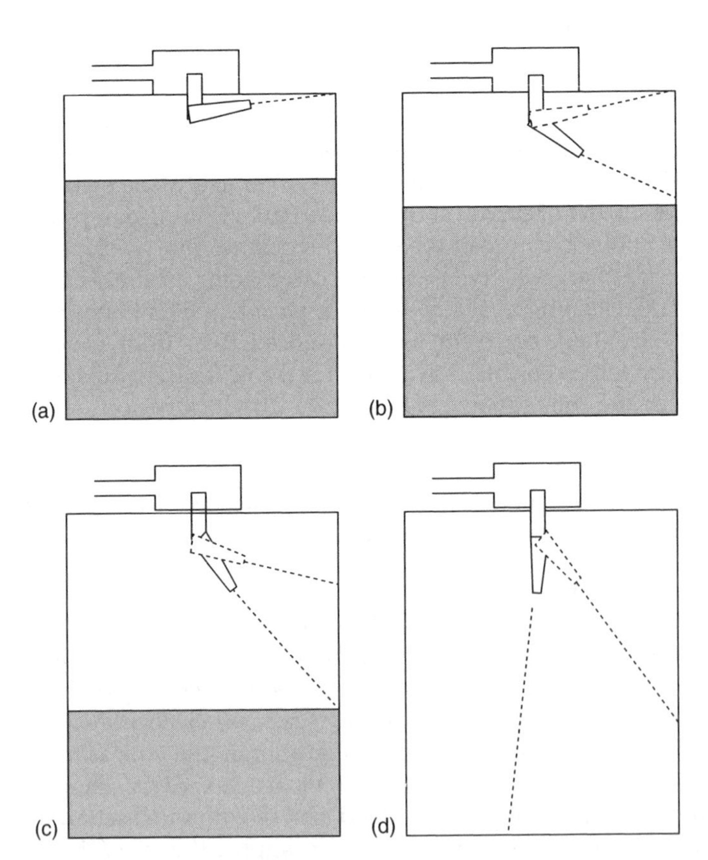

Fig. 5.16 COW cycles. (a) First Cycle Stage One – nozzle elevated for upper level wash. (b) Second Cycle Stage Two – one-third of cargo discharged, nozzle programmed to wash upper third of tank. (c) Third Cycle Stage Three – two-thirds of the cargo is discharged. Nozzle programmed to wash mid levels of the tank. (d) Fourth Cycle, last stage – machine programmed so that the lower levels and the last washing cycle coincide with the end of discharge.

- *Operation – Stage Three*: The third stage is where the machi from where two-thirds of the tank has been discharged an one- and two-thirds of the tanks structure is washed (Figure 5
- *Operation – Last Stage*: The final stage washes the last third and the bottom of the tank with the jet pointing in the downward position (Figure 5.16(d)).

COW – preparation and activities

Prior to arrival at the port of discharge:

1. Has the terminal been notified?
2. Is oxygen-analysing equipment tested and working satisfactorily?
3. Are tanks pressurized with good quality IG (maximum 8% oxygen)?
4. Is the tank-washing pipeline isolated from water heater and engine room?
5. Are all the hydrant valves on the tank-washing line securely shut?
6. Have all tank-cleaning lines been pressurized and leakages made good?

In port:

1. Is the quality of the IG in the tanks satisfactory (8% oxygen or less)?
2. Is the pressure on the IG satisfactory?
3. Have all discharge procedures been followed and ship-to-shore check-list completed?

Before washing:

1. Are valves open to machines on selected tanks for washing?
2. Are responsible persons positioned around the deck to watch for leaks?
3. Are tank ullage gauge floats lifted on respective tanks to be washed?
4. Is the IG system in operation?
5. Are all tanks closed to the outside atmosphere?
6. Have tanks positive IG pressure?

During washing:

1. Are all lines oil tight?
2. Are tank-washing machines functioning correctly?
3. Is the IG in the tanks being retained at a satisfactory quality?
4. Is positive pressure available on the IG system?

After washing:

1. Are all the valves between discharge line and the tank-washing line shut down?
2. Has the tank-washing main pressure been equalized and the line drained?
3. Are all tank-washing machine valves shut?

After departure:

1. Have any tanks due for inspection been purged to below the critical dilution level prior to introducing fresh air?
2. Has oil been drained from the tank-washing lines before opening hydrants to the deck?

The IG system

Tanker vessels have an inherent danger from fire and/or explosion and it is desirable that the atmosphere above an oil cargo or in an empty tank is such that it will not support combustion. The recognized method of achieving this status is to keep these spaces filled with an IG. Such a system serves two main functions:

1. Use of IG inhibits fire or explosion risk
2. It inhibits corrosion inside cargo tanks.

As IG is used to control the atmosphere within the tanks it is useful to know exactly what composition the gases are, not only from a safety point of view but to realize what affect such an atmosphere would have on the construction of the tanks.

Boiler flu gas consists of the following mix (assuming a well-adjusted boiler):

Component	*Percentage of IG*
Nitrogen	83
Carbon dioxide (CO_2)	13
Carbon monoxide	0.3
Oxygen	3.5
Sulphur dioxide	0.005
Nitrogen oxides	Traces
Water vapour	Traces
Ash	Traces
Soot	Traces

Flu gases leave the boiler at about 300°C, contaminated with carbon deposits and sulphurous acid gas. The gas then passes through a scrubber which washes out the impurities and reduces the temperature to within 1°C of the ambient sea temperature.

The clean cooled gas is now moisture laden and passes through a demister where it is dried. It is then fan assisted on passage towards the cargo tanks passing through a deck water seal and then over the top of an oil seal to enter at the top of the tank. It is allowed to circulate and is purged through a pipe which extends from the deck to the bottom of the tank (Figure 5.17).

There is a sampling cock near the deck water seal for monitoring the quality of the IG. Individual tank quality is tested by opening the purge pipe cover and inserting a sample probe.

Excess pressure in the cargo tanks being vented through a pressure vacuum valve (P/V valve) set at 2 psi, which is then led to a mast riser fitted with a gauze screen. The excess is then vented to atmosphere as far from the deck as practicable.

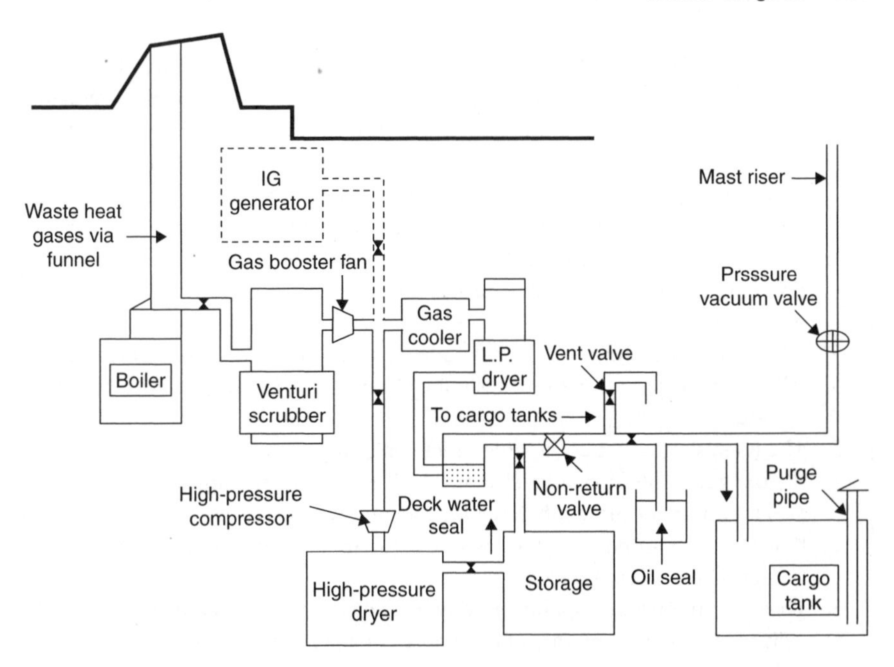

Fig. 5.17 The IG system.

Requirements for IG systems

Additional reference should be made to the Revised Guidelines for Inert Gas Systems adopted by the Maritime Safety Committee, June 1983 (MSC/Circ. 353).

In the case of chemical tankers, reference, Resolution A. 567(14) and A. 473(XIII).

Tankers of 20 000 tonnes deadweight and above, engaged in carrying crude oil, must be fitted with an IG system:

1. Venting systems in cargo tanks must be designed to operate to ensure that neither pressure nor vacuum inside the tanks will exceed design parameters, for volumes of vapour, air or IG mixtures.
2. Venting of small volumes of vapour, air or IG mixtures, caused by thermal variations effecting the cargo tank, must pass through 'P/V valves'.

   Large volumes caused by cargo loading, ballasting or during discharge must not be allowed to exceed design parameters.

   A secondary means of allowing full flow relief of vapour, air or IG mixtures, to avoid excess pressure build-up must be incorporated, with a pressure sensing, monitoring arrangement. This equipment must also provide an alarm facility activated by over-pressure.

3. Tankers with double-hull spaces and double-bottom spaces shall be fitted with connections for air and suitable connections for the supply of IG. Where hull spaces are fitted to the IG permanent distribution system, means must be provided to prevent hydrocarbon gases from cargo tanks, entering double-hull spaces (where spaces are not permanently connected to the IG system appropriate means must be provided to allow connection to the IG main).
4. Suitable portable instruments and/or gas-sampling pipes for measuring flammable vapour concentrations and oxygen must be provided to assess double-hull spaces.
5. All tankers operating with a COW system must be fitted with an IG system.
6. All tankers fitted with an IG system shall be provided with a closed ullage system.
7. The IG system must be capable of inerting empty cargo tanks by reducing the oxygen content to a level which will not support combustion. It must also maintain the atmosphere inside the tank with an oxygen content of less than 8% by volume and at a positive pressure at all times in port or at sea, except when necessary to gas free.
8. The system must be capable of delivering gas to the cargo tanks at a rate of 125% of the maximum rate of discharge capacity of the ship, expressed as a volume.
9. The system should be capable of delivering IG with an oxygen content of not more than 5% by volume in the IG supply main to cargo tanks.
10. Flue gas isolating valves must be fitted to the IG mains, between the boiler uptakes and the flue gas scrubber. Soot blowers will be arranged so as to be denied operation when the corresponding flue gas valve is open.
11. The 'scrubber' and 'blowers' must be arranged and located aft of all cargo tanks, cargo pump rooms and cofferdams separating these spaces from machinery spaces of Category 'A'.
12. Two fuel pumps or one with sufficient spares shall be fitted to the IG generator.
13. Suitable shut offs must be provided to each suction and discharge connection of the blowers. If blowers are to be used for gas freeing they must have blanking arrangements.
14. An additional water seal or other effective means of preventing gas leakage shall be fitted between the flue gas isolating valves and scrubber, or incorporated in the gas entry to the scrubber, for the purpose of permitting safe maintenance procedures.
15. A gas-regulating valve must be fitted in the IG supply main, which is automatically controlled to close at predetermined limits.
    (This valve must be located at the forward bulkhead of the foremost gas-safe space.)
16. At least two non-return devices, one of which will be a water seal must be fitted to the IG supply main. These devices should be located in the cargo area, on deck.
17. The water seal must be protected from freezing, and prevent backflow of hydrocarbon vapours.

18. The second device must be fitted forward of the deck water seal and be of a non-return valve type or equivalent, fitted with positive means of closing.
19. Branch piping of the system to supply IG to respective tanks must be fitted with stop valves or equivalent means of control, for isolating a tank.
20. Arrangements must be provided to connect the system to an external supply of IG.
21. Meters must be fitted in the navigation bridge of combination carriers which indicate the pressure in slop tanks when isolated from the IG main supply. Meters must also be situated in machinery control rooms for the pressure and oxygen content of IG supplied (where a cargo control room is a feature these meters would be fitted in such rooms).
22. Automatic shutdown of IG blowers and the gas-regulating valve shall be arranged on predetermined limits.
23. Alarms shall be fitted to the system and indicated in the machinery space and the cargo control room. These alarms monitor the following:
    - Low water pressure or low water flow rate to the flu gas scrubber.
    - High water level in the flu gas scrubber.
    - High gas temperature.
    - Failure of the IG blowers.
    - Oxygen content in excess of 8% by volume.
    - Failure of the power supply to the automatic control system, regulating valve and sensing/monitoring devices.
    - Low water level in the deck water seal.
    - Gas pressure less than 100-mm water gauge level.
    - High gas pressure.
    - Insufficient fuel oil supply to the IG generator.
    - Power failure to the IG generator.
    - Power failure to the automatic control of the IG generator.

Hazards with IG systems

The IG system aboard any vessel has two inherent hazards:

1. If the cooling water in the scrubber should fail, then uncooled gas at 300°C would pass directly to the cargo tank. This is prevented by the fitting of two water sensors in the base of the scrubber which, if allowed to become uncovered, would generate an alarm signal which shuts the system down and vents the gas to atmosphere. In the event that both sensors failed two thermometer probes at the outlet of the scrubber would sense an unacceptable rise in temperature and initiate the same shutdown procedure.
2. If there was a failure in the P/V valve, at the same time as a rise in the pressure within the cargo tank, it would result in pressure working backwards towards the boiler with a possible risk of explosion. This is prevented by the water in the deck seal forming a plug in the IG line until

a sufficient head is generated to blow out the oil seal and the excess pressure vents to the deck. The pressure of water in the water seal is essential; therefore, the two water sensors would sense its absence and shut down the plant as previously stated.

IG pressure should be maintained at a positive pressure at all times, to avoid air being forced into the cargo spaces. Such a positive pressure is also exerted onto the surface of the oil cargo and assists in pushing the oil along the suction line towards the cargo pump, and in so doing assists the draining of the tanks. Any excess pressure in the cargo tanks is vented through the P/V valve.

IG – voyage cycle

- *Phase 1* – Vessel departs dry dock with all tanks vented to atmosphere and partially ballasted. The IG plant is started, empty tanks and ullage spaces purged to atmosphere until oxygen levels are acceptable. IG quality should be monitored and maintained throughout the ballast voyage.
- *Phase 2* – Prior to arrival at the loading port the IG plant would be started and ballast reduced to about 25% of the ships deadweight, ballast being replaced by IG. After berthing, the remainder of the seawater ballast would be discharged and replaced by IG. The IG plant would then be shut down, the deck isolation valve would be closed and the mast riser opened, prior to commencing loading. IG would be displaced through mast risers. On completion of loading, the IG would be topped up to a working pressure which would be maintained though the loaded voyage (this would be expected to reduce evaporation and prevent oxygen access).
- *Phase 3* – On arrival at the port of discharge, the IG plant would be set to maximum output with discharge pumps at maximum output. The IG pressure should be monitored carefully and if it approaches a negative, the rate of discharge of the cargo reduces. The mast riser must never be opened to relieve the vacuum during the discharge period.
- *Phase 4* – On completion of discharge, the IG system should be shut down. If and when ballasting takes place the IG and hydrocarbons would be vented to atmosphere.
- *Phase 5* – On departure from the discharge port all tanks must be drained to the internal slop tank, then purged with IG to reduce the hydrocarbon levels to below 2%.
- *Phase 6* – Tank cleaning can now be permitted with IG in fully inerted tanks. This weakens the hydrocarbon level and the positive pressure prevents pumps draining or drawing atmosphere into the tanks.
- *Phase 7* – When all the vessels tanks have been washed and ballast changed it may be necessary to carry out tank inspections. If this is the case, all tanks would then have to be purged with IG to remove all traces of hydrocarbon gas before venting by fans. All tanks would then be tested with explosi-meter and oxygen analyser (full procedure for enclosed space entry must be observed before internal inspection).

Advantages and disadvantages of the IG system

Advantages

1. A safe tank atmosphere is achieved which is non-explosive
2. It allows high-pressure tank washing and reduces tank-cleaning time
3. It allows COW
4. Reduces corrosion in tanks – with an efficient scrubber in the system
5. Improves stripping efficiency and reduces discharge time
6. Aids the safe gas freeing of tanks
7. It is economical to operate
8. It forms a readily available extinguishing agent for other spaces
9. Reduces the loss of cargo through evaporation
10. Complies with legislation and reduces insurance premiums.

Disadvantages

1. Additional costs for installation
2. Maintenance costs are incurred
3. Low visibility inside tanks
4. With low oxygen content, tank access is denied
5. Could lead to contamination of high-grade products
6. Moisture and sulphur content corrodes equipment
7. An established reverse route for cargo to enter the engine room
8. Oxygen content must be monitored and alarm sensed at all times
9. Instrumentation failure could affect fail-safe devices putting the ship at risk through the IG system
10. An additional gas generator is required in the system in the absence of waste heat products from boiler flue gases.

> **Note:** *Instrumentation of the system to cover: IG temperature pressure read outs and recorders. Alarms for: blower failure, high oxygen content alarm, high and low gas pressure alarms, high gas temperature, low seawater pressure and low level alarm in the scrubber and the deck water seal, respectively.*

Deck water seal operation

The water level in the deck water seal is maintained by constant running of the seawater pump and a gooseneck drain system. Under normal IG pressure the IG will bubble through the liquid from the bottom of the IG inlet pipe and exit under normal operating pressure. In the event of a back pressure developing and the water surface experiencing increased pressure, this would force the water level up the IG inlet pipe, sealing this pipe entrance and preventing hydrocarbons entering the scrubber (Figure 5.18).

Tank atmosphere

The Cargo Officer will need to be able to assess the condition of the atmosphere inside the tank on numerous occasions. To this end, various monitoring

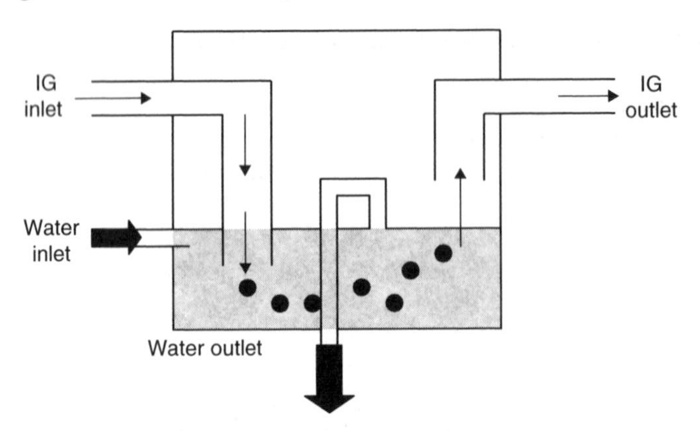

Fig. 5.18 Deck water seal operation.

equipment is available to carry out 'gas detection' and 'oxygen content'. The officer should be familiar with the type of equipment aboard his/her own vessel and have a degree of understanding how such instrumentation operates.

Gas detection

It should be understood from the outset that many accidents and loss of life has occurred through lack of knowledge of gas-detection methods and the correct practice concerning this topic. The explosi-meter, of which there are several trade names available, is used for detecting the presence of flammable gas and/or air mixture.

The explosi-meter The explosi-meter is an instrument which is specifically designed for measuring the lower flammable limit (LFL). It will only function correctly if the filament has an explosive mixture in contact with it. It is contained in a hand-held size box with a battery power supply (Figure 5.19).

When in use, the sample tube is lowered into the tank and a sample of the atmosphere is drawn up into the instrument by several depressions of the rubber aspirator bulb. If the sample contains an explosive mixture the resistance of the catalytic filament will change due to the generated heat. An imbalance of the wheat-stone bridge is detected by the ohmmeter which tells the operator that hydrocarbon gas is present in the tank in sufficient quantity to support combustion.

Note: Combustibles in the sample are burned on the heated filament, which raises its temperature and increases the resistance in proportion to the concentration of combustibles in the sample. This then causes the imbalance in the wheat-stone bridge.

However, it should be realized that a zero reading does not necessarily indicate that there is no hydrocarbon gas present, nor does it mean that no oxygen is present. All it signifies is that the sample taken is either too rich or too lean to support combustion. Care must be taken when testing the

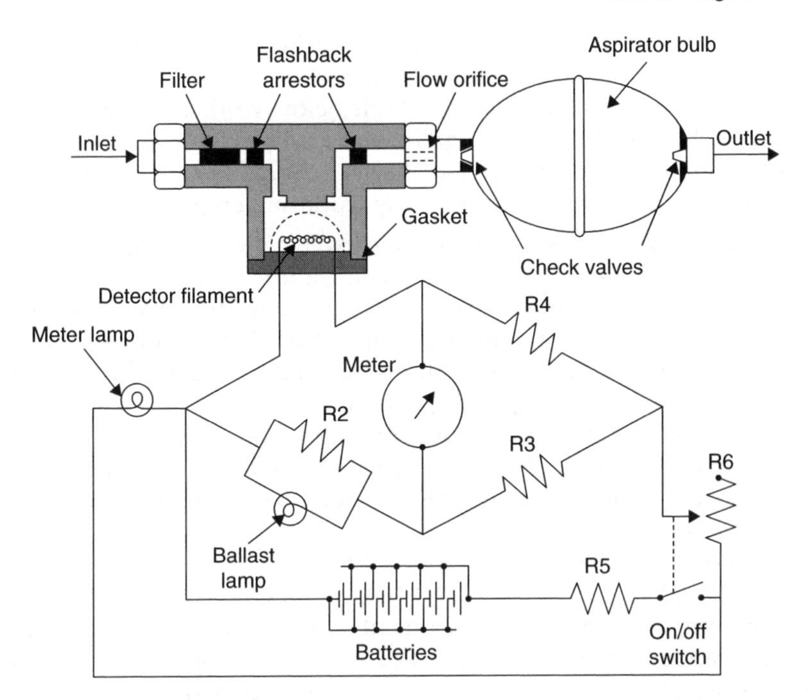

Fig. 5.19 The MSA model 2E explosi-meter (combustible gas detector).

atmosphere in enclosed spaces to give consideration for the relative vapour density where mixtures of gases are encountered. A test at one particular level in a tank should be realized as not necessarily being an equivalent reading for other different levels in the same tank.

The electrical bridge circuit of the instrument is designed so that its balance is established at the proper operating temperature of the detecting filament. The circuit balance and detector current are adjusted simultaneously by adjustment of the rheostat. The proper relationship between these two factors is maintained by a special ballast lamp in the circuit.

The graduations on the meter are a per cent of the lower explosive limit (LEL) reading between 0% and 100%. A deflection of the meter between 0% and 100% shows how close the atmosphere being tested approaches the minimum concentration required for explosion. When a test is made with the explosi-meter, and a deflection to the extreme right-hand side of the scale is noted and remains there, then the atmosphere under test is explosive.

Limitations of explosi-meters – The explosi-meter has been designed to detect the presence of flammable gases and vapours. The instrument will indicate in a general way whether or not the atmosphere is dangerous from a flammability point of view. It is important to realize that such information obtained from the instrument is appraised by a person skilled in the interpretation of the reading, bearing in mind the environment. For example, the atmosphere sample which is indicated as being non-hazardous from the standpoint of fire and explosion, may if inhaled, be toxic to workers who are exposed to that same atmosphere.

Additionally, a tank that is deemed safe before work is commenced may be rendered unsafe by future ongoing operations, e.g. stirring or handling bottom sludge. This would indicate the need for regular testing practices to be in place in questionable spaces while work is in progress.

Explosi-meter special uses – Where the explosi-meter is employed to test an atmosphere which is associated with high boiling point solvents, it should be borne in mind that the accuracy of the reading may be questionable. The space may be at a higher temperature than the instrument, and therefore it must be anticipated that some condensation of combustible vapours would be in the sampling line. As a consequence, the instrument could read less than the true vapour concentration.

A way around this would possibly be to warm the sampling line and the instrument unit to an equivalent temperature as that of the space being tested.

> ***Note:*** *Under no circumstances should such instruments be heated over 65°C (150°F).*

Furthermore, some types of instruments are designed to measure combustible vapours in air. They are not capable of measuring the percentage vapours in a steam or inert atmosphere, due to the absence of oxygen necessary to cause combustion.

Care in use – When sampling over liquids, care should be taken that the sampling tube does not come into contact with the liquid itself. A probe tube can be used in tests of this character, to prevent liquid being drawn into the sampling tube.

Drager instruments This is an instrument which draws a gas or vapour through an appropriate glass testing tube, each tube being treated with a chemical that will react with a particular gas, causing discolouration progressively down the length of the tube. When measured against a scale, the parts per million (ppm) can be ascertained.

The instrument is used extensively on the chemical carrier trades though it does have tubes for use with hydrocarbons, which make it suitable for use on tankers.

Alarm system detectors An instrument which is taken into a supposedly gas-free compartment and used while work is ongoing. If gas is released or disturbed in the work place a sensitive element on the instrument triggers an audible and visual alarm. Once the alarm has been activated personnel would be expected to evacuate the compartment immediately.

The oxygen analyser

In order for an atmosphere to support human life it must have the oxygen content of 21%. The oxygen analyser is an instrument that measures the oxygen content of an atmosphere to establish whether entry is possible, but it is also employed for inerted spaces which must be retained under 5% oxygen to affect a safe atmosphere within the tank (Figure 5.20).

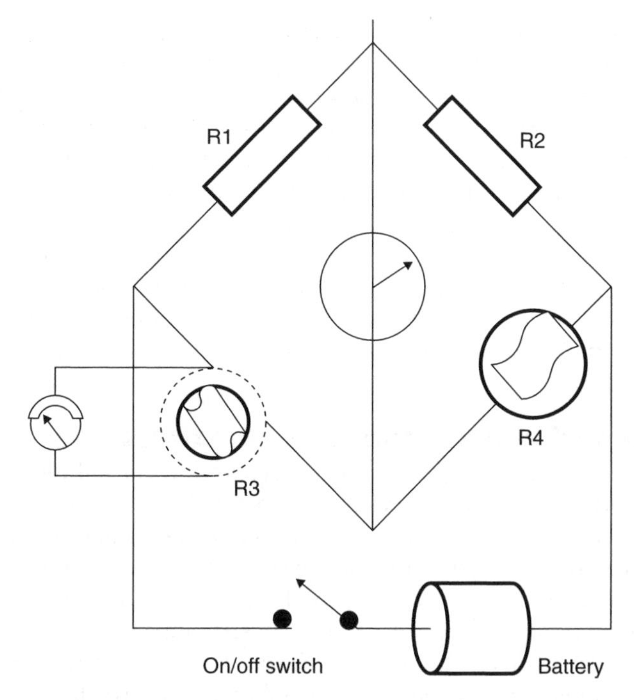

Fig. 5.20 Oxygen analyser – circuitry principle.

The oxygen sensor will be either an electromagnetic heated filament or an electrochemical resistor cell. The instrument was designed to measure the oxygen content only and will not detect the presence of any other gases. As shown in Figure 5.20, the resistor filaments R3 and R4 are of equal rating. The resistor filament R3 is surrounded by a magnetic field. The atmosphere sample drawn past the filament will depend on the permitted current flow through the coil and meter, depending on the amount of oxygen in the sample.

Oxygen analysers are portable instruments which draw a sample of the atmosphere for testing through a sampling hose by means of a rubber aspirator bulb. The principle of operation is a self-generating electrolytic cell in which the electric current is directly proportional to the percentage oxygen in a salt solution connecting to the electrodes. The electrodes are connected to a micro-ammeter, so that the current read by the meter can be calibrated to indicate directly the percentage oxygen of the sample.

There are variations and different types of instruments available. Manufacturer's instructions and manuals for use and maintenance should therefore be followed when these instruments are employed.

Chemical reaction measuring device

Gas detection can also be achieved by using a test sample of the atmosphere to pass over a chemical-impregnated paper or crystal compound. The chemicals subsequently react with specific gases on contact.

The amount of discolouration occurring in the crystals or on the paper can then be compared against a scale to provide the amount of gas within the sample. The operation uses a bellows to draw through a 100 cm^3 of sample gas and a variety of tubes can be used to indicate specific gases. Example gases indicated are likely to be, but not limited to carbon monoxide, hydrogen sulphide, hydrocarbon, radon, nitrous oxide. A popular instrument is the 'Drager Tube System' for gas detecting.

Although well-used in the industry, the system does have drawbacks in the fact that the tubes required for different gases have a limited shelf life. The bellows can develop leaks and they can be affected by temperature extremes. Tube insertion must also be carried out the correct way.

Coastal and shuttle tanker operations

Numerous small tanker operations are engaged in coastal regions around the world and employ the services of coastal-sized craft to shuttle cargo parcels between main terminal ports, FSUs and the smaller out of the way ports. Restrictions are often put on direct delivery from the ocean-going vessels because of the available depth of water in the smaller enclaves and as such the geography imposes draught restrictions on the larger vessels. This particular drawback is also affecting the container trade, with container vessels currently being increased in overall size, the larger vessels are finding some ports are not available to them because of similar draught restrictions (Figure 5.21).

Fig. 5.21 The coastal oil tanker 'Alacrity' lies port side to a terminal berth in the UK.

Examples of tanker cargoes

Bitumen – this cargo solidifies at normal temperatures and must be kept hot during transit. Ships are specifically designed for this trade, with large centre tanks and additional heating coils. The centre tanks being used for cargo and the wing tanks for ballast.

Chemicals (various) – precautions for these cargoes as outlined in previous text. Additional reference to the International Maritime Dangerous Goods (IMDG) Code and respective precautions pertaining to the type of commodity.

Creosote – this is a very heavy cargo and requires constant heating during the voyage.

Crude oil – varies greatly with RD and viscosity. It is not heated unless a very heavy grade, as heating evaporates the lighter fractions. Crude oil has a high fire risk.

Diesel oil – is an intermediate between fuel oil and gas oil. It is generally regarded as a dirty oil but its viscosity is such that it does not require heating prior to discharge.

Fuel oil – is a black oil and is graded according to its weight and viscosity. It has a low fire risk and generally requires heating prior to discharge.

Gas oil – this is a clean oil and is used for light diesel engines as well as for making gas. A reasonable level of cleanliness is required before loading this cargo, which may be used as a transition cargo when a ship is being changed from a black oil carrier to a clean oil trade. Fire risk is low and no heating is required.

Gasoline (petrol) – is light and volatile. It has a high fire risk and may easily be contaminated if loaded into tanks which are not sufficiently clean.

Grain – may be successfully carried in selective tankers since when in a bulk state it has many of the qualities of a liquid. It requires very careful tank preparation and tankers would only normally enter the trade if the oil market was depressed.

Kerosene (paraffin) – this is a clean oil which is easily discoloured. Precautions should be taken to prevent the build-up of static. These may include a slow loading and discharge pattern being employed.

Latex – an occasional cargo carried in tankers and in ships 'deep tanks'. Usually has added ammonia. The tanks should be exceptionally clean and fitted with pressure relief valves. Steelwork is pre-coated in paraffin wax and heating coils in tanks should be removed. Following discharge the tanks should be washed with water to remove all traces of ammonia.

Liquefied gases – generally carried in specifically designed vessels for the transport of LNG and liquid propane gas (a liquefied petroleum gas, LPG).

Lubricating oils – these are valuable cargoes and are usually shipped in the smaller product carriers. Good separation is necessary to avoid contamination between grades. Tanks and pipelines must be free of water before loading. Some grades may require heating before discharging.

Molasses – a heavy viscous cargo which is normally carried in designated tankers specific for the trade. A comprehensive heating system is necessary and special pumps are provided to handle the thick liquid.

Propane – a gas similar to 'butane', see liquefied gases.

Vegetable oils – these are generally carried in small quantities in the deep tanks of cargo ships but some such as 'linseed oil' may be carried in tankers. Exceptional cleanliness of the tanks is required prior to loading such a cargo.

Whale oil – whale factory ships are basically tankers carrying fuel oil on the outward passage and whale oil when homeward bound. Careful cleaning is required before carrying whale oil in tanks which previously carried fuel oil (in recent years the practice of whale hunting has been severely restricted).

Wine – can be carried in tankers but they are usually dedicated ships to the trade. Similar vessels sometimes engage in the carriage of fruit juices, especially orange juice. A high degree of cleanliness in the tanks is expected (Ref. page 162).

Product tankers

Product tankers tend to be smaller and more specialized than the large crude oil carriers and generally lay alongside specialized berths when loading and discharging, employing specialist product lines to avoid contamination of cargoes (Figures 5.22 and 5.23).

Fig. 5.22 Conventional large oil tanker seen passing through the Dardanelles.

Fig. 5.23 The product tanker 'Folesandros' lies port side to the berth, discharging in Gibraltar.

Bulk liquid chemical carriers

Phrases and terminology associated with the chemical industry

Adiabatic expansion – is an increase in volume without a change in temperature or without any heat transfer taking place.

Anaesthetics – chemicals that affect the nervous system and cause anaesthesia.

Aqueous – a compound within a water-based solution.

Auto-ignition – a chemical reaction of a compound causing combustion without a secondary source of ignition.

Boiling point – that temperature at which a liquid's vapour pressure is equal to the atmospheric pressure.

Catalyst – a substance that will cause a reaction with another substance or one that accelerates or decelerates a reaction.

Critical pressure – that minimum pressure which is required to liquefy a gas at its critical temperature.

Critical temperature – that maximum temperature of a gas at which it can be turned into a liquid by pressurization.

Filling ratio – that percentage volume of a tank which can be safely filled allowing for the expansion of the product.

Freezing point – that temperature at which a substance must be at to change from a liquid to a solid state or vice versa.

Hydrolysis – that process of splitting a compound into two parts by the agency of water. One part being combined with hydrogen, the other with hydroxyl.

Hydroscopic – that ability of a substance to absorb water or moisture from the atmosphere.

Inhibitor – a substance which, when introduced to another, will prevent a reaction.

Narcosis – a human state of insensibility resembling sleep or unconsciousness, from which it is difficult to arouse.

Oxidizing agent – an element or compound that is capable of adding oxygen to another.

Padding – a procedure of displacing air or unwanted gasses from tanks and pipelines with another compatible substance, e.g. IG, cargo vapour or liquid.

Polymerization – that process which is due to a chemical reaction within a substance, capable of changing the molecular structure within that substance, i.e. liquid to solid.

Reducing agent – an element or compound that is capable of removing oxygen from a substance.

Reid vapour pressure – is that vapour pressure of a liquid as measured in a Reid apparatus at a temperature of 100°F expressed in psi/°A.

Self-reaction – is that ability of a chemical to react without other influence which results in polymerization or decomposition.

Sublimation – that process of conversion from a solid to a gas, without melting (an indication that the flash point is well below the freezing point).

Threshold limit value – is that value reflecting the amount of gas, vapour, mist or spray mixture that a person may be daily subjected to, without suffering any adverse effects (usually expressed in ppm).

Vapour density – that weight of a specific volume of gas compared to an equal volume of air, in standard conditions of temperature and pressure.

Vapour pressure – that pressure exerted by a vapour above the surface of a liquid at a certain temperature (measured in mm of mercury, mmHg).

Bulk chemical cargoes

The term liquid chemicals within the industry is meant to express those chemicals in liquid form at an ambient temperature or which can be liquefied by heating, when carried at pressures up to 0.7 kg/cm^2. Above this the chemical pressure would fall into the category of 'liquefied gases'.

The chemicals carried at sea have a variety of properties. Nearly half of the 200 chemicals commonly carried have fire or health hazards no greater than petroleum cargoes. Therefore, they can be safely carried by way of ordinary product carriers, though some modification is sometimes required to avoid contamination.

Other chemical substances require quality control much more stringent than petroleum products. Contamination, however slight, cannot be allowed to occur and for this reason tanks are nearly always coated or made of special materials like stainless steel.

Extreme care must be exercised when loading such cargoes that any substance which could cause a reaction are kept well separated. To ensure quality and safe carriage, separate pipelines, valves and separate pumps are the norm for specific cargo parcels. Also reactionary chemicals cannot be placed in adjacent tanks with only a single bulkhead separation. Neither can pipelines carrying one substance pass through a tank carrying another substance with which it may react. Chemical products which react with sea water are carried in centre tanks while the wing tanks are employed to act as cofferdams.

Chemical carriers require experienced and specialized trained personnel in order to conduct their day-to-day operations safely. They also require sophisticated cargo-handling and monitoring equipment. The ships must conform in design and construction practice to the IMO 'Code for the Construction and Equipment of Ships Carrying Dangerous Chemicals in Bulk'.

The purpose of the 'code' is to recommend suitable design criteria, safety measures and construction standards for ships carrying dangerous chemical substances. Much of the content of the code has been incorporated into the construction regulations produced by the Classification Societies.

Classification – chemical carriers

(Chapter references to the International Code for the Construction and Equipment of Ships Carrying Dangerous Chemicals in Bulk) (IBC)

In general ships carrying chemicals in bulk are classed into three types:

1. *A 'Type 1' ship* is a chemical tanker intended to transport Chapter 17 of the IBC Code products with very severe environmental and safety hazards which require maximum preventive measures to preclude an escape of such cargo.
2. *A 'Type 2' ship* is a chemical tanker intended to transport Chapter 17 of the IBC Code products with appreciably severe environmental and safety hazards which require significant preventive measures to preclude an escape of such cargo.

3. *A 'Type 3' ship* is a chemical tanker intended to transport Chapter 17 of the IBC Code products with sufficiently severe environmental and safety hazards which require a moderate degree of containment to increase survival capability in a damaged condition.

Many of the cargoes carried in these ships must be considered as extremely dangerous and, as such, the structure of the ship's hull is considered in the light of the potential danger, which might result from damage to the transport vessel. Type 3 ships are similar to product tankers in that they have double hulls but have a greater subdivision requirement. Whereas Types 1 and 2 ships must have their cargo tanks located at specific distances inboard to reduce the possibility of impact load directly onto the cargo tank (Figure 5.24).

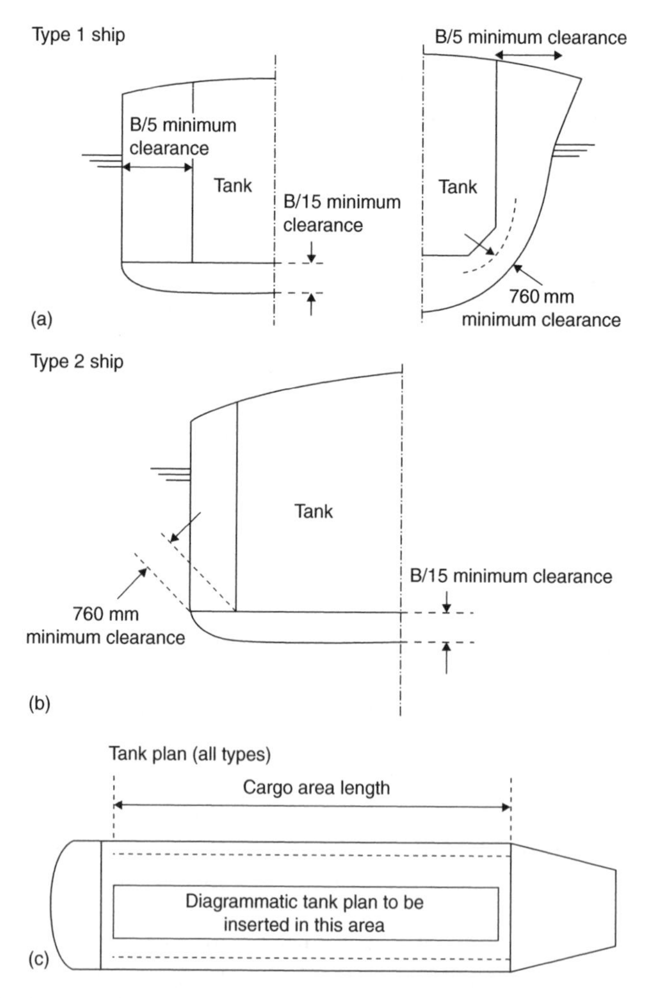

Fig. 5.24 The tank arrangement must be attached to the International Certificate of Fitness, for the carriage of Dangerous Chemicals in Bulk.

Parcel tankers – construction features

Ships built specifically as parcel tankers with the intention of carrying a wide variety of cargoes will generally have some tanks of 'stainless steel' or tanks clad in stainless steel. For reasons of construction and cost this means having a double skin. Mild steel tanks may similarly be built with side cofferdams and a double bottom and are usually coated in either epoxy or silicate. Chemicals of high density like 'ethylene dibromide' may have specially constructed tanks or in some cases only carry partly filled cargo tanks (Figure 5.25).

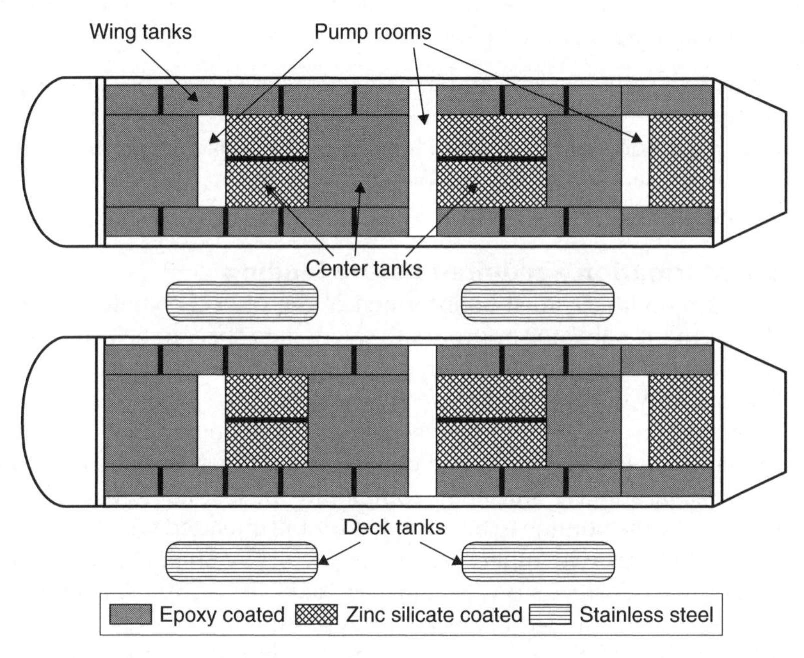

Fig. 5.25 Diagram of a parcel tanker.

Similarly, cargoes with higher vapour pressures may generate a need for tanks to be constructed to withstand higher pressures than say the conventional tanker – particularly relevant where the boiling point of the more volatile cargoes is raised and the risk of loss is increased.

The IBC Code specifies requirements for safety equipment to monitor vapour detection, fire protection, ventilation in cargo-handling spaces, gauging and tank filling. Once all criteria is met the Marine Authority (Maritime and Coastguard Agency (MCA) in the UK) will issue, on application an MCA/IMO Certificate of Fitness for the Carriage of Dangerous Cargoes in Bulk.

Vapour lines

In general, each tank will have its own vapour line fitted with P/V valves but grouped tanks may have a common line. Since some vapours from specific

ghly toxic or flammable, the lines are led well over accommo-
e expected to release vapour as near as possible in a vertical
e vessels carry provision to return vapour expelled during
ocess to the shoreside tank. Examples are when the cargo is
highly toxic or the chemicals react dangerously with air.

Main hazards associated with chemicals to humans

The substances carried in chemical tankers present certain hazards to operations of transport and to the crews of the ships. The main hazards fall into one of a combination of the following:

1. danger to health – toxicity and irritant characteristics of the substance or vapour
2. water pollution aspect – human toxicity of the substance in the solution
3. reactionary activity with water or other chemicals
4. fire and/or explosion hazard.

Cargo information – required before loading

1. The Cargo Officer must be informed of the correct chemical name of the cargo to enable the appropriate safety data sheet to be consulted in the Tanker Safety Guide (Chemicals).
2. The quantity of cargo and respective weight.
3. Clearance on quality control must be confirmed. Contamination, usually measured in ppm so tanks and pipelines must be assured to be clean.
4. The specific gravity value of the commodity must be advised to allow an estimate of the volume to be occupied for the intended weight of cargo.
5. Incompatibility with other cargoes or specifically other chemicals must be notified. Correct stowage must be achieved so that incompatible cargoes are not stowed in adjacent compartments.
6. Temperature of the cargo: (a) at the loading stage and (b) during the carriage stage. This criterion is required because temperature of the commodity will affect the volume of the total cargo loaded, while the expected carriage temperature will indicate whether heating of the cargo will be required.
7. The tank-coating compatibility must be suitable for the respective cargo.
8. Any corrosive properties of the chemical. This information would also be relevant to the tank-coating aspect and provide possible concerns for incurring damage to shipboard fittings.
9. Electrostatic properties can be acquired by some chemicals. With this in mind the principles applicable to hydrocarbons should be applied.
10. Data on the possibility of fire or explosion – 50% of chemicals carried are derived from hydrocarbons and the risk of fire or explosion is similar to the carriage for hydrocarbons.
11. The level of toxicity of the chemical. If high-toxic vapours are a characteristic of the cargo then enclosed ventilation may be a requirement.
12. Health hazards of any particular parcel of cargo.

13. Reactivity with water, air or other commodities.
14. What emergency procedures must be applicable in the event of contact or spillage.

The chemical data sheets of respective cargoes usually provide all of the above along with additional essential information for the safe handling and carriage of the commodity.

The protection of personnel

The hazards of the chemical trade have long been recognized and the need for personal protection of individuals engaged on such ships must be considered as the highest of priorities. A chemical cargo can be corrosive and destroy human tissue on contact. It can also be poisonous and can enter the body by several methods. It may be toxic and if inhaled damage the brain, the nervous system or the body's vital organs. Additionally, the chemical may give off a flammable gas giving a high risk of fire and explosion.

The IMO (IBC) Code requires that personnel involved in cargo operations aboard chemical carriers be provided with suitable protection by way of clothing and equipment, which will give total coverage of the skin in a manner that no part of the body is left exposed (i.e. chemical suits).

Protective equipment

Protective equipment to include:

1. full protective suit manufactured in a resistant material with tight-fitting cuff and ankle design
2. protective helmet
3. suitable boots
4. suitable gloves
5. a face shield or goggle protection
6. a large apron.

Where the product has inhalation problems for individuals then the above equipment would be supplemented by breathing apparatus (B/A).

Where toxic cargoes are carried, SOLAS requires that the ship should carry a minimum of three (3) *additional* complete sets of safety equipment, over and above the SOLAS '74 requirements.

Safety equipment set

Safety equipment set shall comprise:

1. a self-contained B/A (SCBA)
2. protective clothing (as described above)
3. steel core rescue line and harness
4. explosive-proof safety lamp.

An air compressor, together with spare cylinders, must also be carried and all compressed air equipment must be inspected on a monthly basis and tested annually.

Where toxic chemical products are carried, all personnel on board the vessel must have respiratory equipment available. This equipment must have adequate endurance to permit personnel to escape from the ship in the event of a major accident.

Associated operations

Heating of cargoes

Certain cargoes are required to be carried and/or discharged at high temperatures and to this end, heating while inside the ship's tanks must take place. Heating is usually provided by either heating coils inside the tanks themselves or, in the case of double-hull vessels, by heating channels on the outside of the tanks. The medium used is either steam, hot water or oil, but care must be taken that the medium is compatible with the cargo.

Tanks that contain chemicals, which could react with each other, must not be on the same heating circuit. Another safety factor is that a heat exchanger must be used between the boiler and the cargo system. This would prevent the possibility of the cargo product finding its way into the ship's boilers, in the event of a leak occurring in the system.

IG systems with chemical cargoes

IG, usually nitrogen, is used to blanket some cargoes. These are usually ones that react with air or water vapour in the atmosphere. They are loaded into tanks after they have been purged with IG and the tank must remain inerted until cleaning has been completed. Other cargoes have the ullage space inerted either as a fire precaution or to prevent reactions, which, while not necessarily dangerous, may put the cargo off specification. The nitrogen is supplied by a shipboard generator or from ashore, or from storage cylinders.

Precautions during loading, discharging and tank cleaning

In addition to the usual safety precautions for tanker practice, if handling toxic cargoes full protective clothing, including B/A, should be worn by all persons on deck. Goggles should be worn when handling cargoes which may cause irritation to the eyes. Such vessels are generally equipped with decontamination deck showers together with escape sets for each crew member.

Tank cleaning After discharge of the majority of cargoes, the tanks can be washed out with salt water as a first wash, then finished with a fresh water wash. Stainless steel tanks are usually washed only with fresh water because of damage, which may be incurred to the steelwork by use of sea water. Washing is often assisted by one of a range of cleansing compounds, which can be sprayed onto the tank sides and then washed off. One of the advantages of double-hull construction is that all the stiffening members of the tanks are on the outside of the tank and cleaning and drainage is therefore much easier.

Some special chemicals may require special cleaning procedures and solvent use, and extreme care should be taken that mixtures created are not of a dangerous nature. Similarly, if washing into a slop tank a dangerous mixture of unknown chemical properties should not be generated.

Fire fighting

Fire-fighting arrangements are similar to that aboard petroleum tankers, with the exception that nitrogen is commonly employed as a smothering agent because some cargoes would be incompatible with CO_2. Ships are therefore generally supplied with an adequate supply of nitrogen. Ordinary foam breaks down when used on water-soluble chemicals so a special alcohol-foam is required – so named as being suitable for fires involving alcohol. In addition, large fixed dry powder plants may be provided for use on the tank deck. Some specialized cargoes require specific fire-fighting techniques and relevant details can be obtained from the shore authorities, prior to loading.

> ***Note:*** *Many cargoes give off harmful vapours when burning and fire parties are advised to ensure that they wear protective clothing and B/A when fighting chemical fires.*

Compatibility

Great care must be taken during the cargo planning stage to ensure that chemicals that react with one another do not come into contact. Such planning is often a shore-based operation which is checked by the Ship's Master or the Chief Officer prior to the commencement of loading.

Chemicals must be located in an appropriate tank according to the IMO Code, and at the same time be compatible with the tank coating as specified in the tables provided by the tank-coating manufacturers. Incompatible cargoes must have positive segregation, and failure to observe such requirements could give rise to a most hazardous situation involving toxics or flammable gas being given off as a by-product.

Additionally, some mixtures of chemicals may react together, but equally some are potentially dangerous on their own. Those that react with air can be contained by IG, or provided with vapour return lines as previously described. However, some react with water (e.g. 'sulphuric acid') and must be loaded in double-skin tanks.

A number of chemicals are self-reactive, in the sense that they may polymerize with explosive violence or cause a generation of considerable heat. Examples of these are 'vinyl acetate' or 'styrene monomers'. If shipped, these have an inhibitor added, but care must be taken with all monomers to ensure that no impurities are introduced, which may act as a catalyst and cause polymerization. Accidental heating with such cargoes should also be avoided.

Volatile cargoes

Such cargoes of a volatile nature must not be stowed adjacent to heated cargoes. The possibility of flammable or toxic vapour release could lead to an after affect which could lead to disastrous consequences should the vapour reach the deck area.

Cargo-handling reference

Most shipping companies have prepared their own operational and safety manuals but most are based on the International Chamber of Shipping (ICS) Tanker Safety Guide for Oil Tankers and Terminals (ISGOTT). This contains an index of chemical names, including synonyms. Cargo information from data sheets for the most common chemicals is also included. Checklists are now also commonly employed to ensure correct procedures are observed throughout all cargo operations.

Merchant Shipping Notices

Merchant Shipping Notices stress the danger from asphyxiation and/or affects of toxic or other harmful vapours. They also strongly advise on the entry procedures into tanks and enclosed spaces alongside the Code of Safe Working Practice (CSWP). Notices emphasize the need for continuous monitoring of the vapour with gas detectors and the necessity of providing adequate ventilation when personnel enter enclosed spaces. Full procedures must include the use of a stand by man at the entrance of an enclosed space while personnel are inside that space.

Compatibility tables

There are various compatibility tables available, but perhaps the most widely applied are the USCG – Bulk Liquid Cargoes Guide to the Compatibility of Chemicals. A hazardous reaction is defined as a binary mixture which produces a temperature rise greater than 25°C or causes a gas to evolve.

The cargo groups for the two chemicals under consideration are first established from an alphabetical listing, then cross-referenced in the compatibility table; unsafe combinations being indicated by an 'X', and reactivity deviations within the chemical groups by the letters 'A' to 'I'.

IMO/IBC code

The International Code for the Construction and Equipment of Ships Carrying Dangerous Chemicals in Bulk and Index of Dangerous Chemicals Carried in Bulk are clearly the main recognized authority regarding the bulk chemical trade. It is recognized as the definitive source of names for products subject to Appendices II and III of Annex II of MARPOL 73/78.

IMO/IGC code

The International Code for the Construction and Equipment of Ships Carrying Liquefied Gases in Bulk. Applicable to all ships regardless of size, inclusive of those vessels under 500-tonne gross, which are engaged in the carriage of liquefied gases having a vapour pressure exceeding 2.8 bar absolute temperature of 37.8°C, and other products as appropriate under Chapter 19 (of the code), when carried in bulk.

(Exception: vessels constructed before October 1994 to comply with Resolution MSC. 5(48) adopted on 17 June 1983).

Bulk liquefied gas cargoes

The liquefied gases which are normally carried in bulk are hydrocarbon gases used as fuels or as feed stocks for chemical processing and chemical gases used as intermediates in the production of fertilizers, explosives, plastics or synthetics. The more common gases are LPGs, such as propane, butane, propylene, butylene, anhydrous ammonia, ethylene, vinyl chloride monomer (VCM) and butadiene. LNG is also transported extensively in dedicated ships (Figure 5.26), LNG being a mixture of methane, ethane, propane and butane with methane as the main component.

Gas properties

Liquefied gases are vapours at normal ambient temperatures and pressures. The atmospheric boiling points of the common gases are given as follows:

LPG
- Propane −42.3°C
- Butane −0.5°C
- LPG propylene −47.7°C
- Butylene −6.1°C

Fig. 5.26 Two LNG carriers lie alongside each other outside the Dubai Dry Dock Complex. These dedicated ships are prominent by the conspicuous cargo domes covering the gas tanks.

Ammonia −33.4°C
Ethylene −103.9°C
VCM −13.8°C
LNG −161.5°C
Butadiene −5.0°C

The carriage of gases in the liquid phase can only be achieved by lowering the temperature or increasing the pressure or a combination of both low temperatures and increased pressures.

The carriage condition is classified as either: 'fully refrigerated' (at approximately atmospheric pressure) or 'semi-refrigerated' (at approximately 0 to −10°C and medium pressure) and fully pressurized (at ambient temperature and high pressures).

LNG and ethylene are normally always carried in the fully refrigerated condition – they cannot be liquefied by increasing the pressure alone – while the LPGs, ammonia, VCM and butadiene can be liquefied by lowering the temperature or increasing the pressure. This permits them to be carried in the fully refrigerated, or the semi-refrigerated or the fully pressurized condition. The IMO/IGC Code provides standards for 'gas tankers' and identifies the types of tanks which must be employed for the carriage of liquefied gases (Figure 5.27).

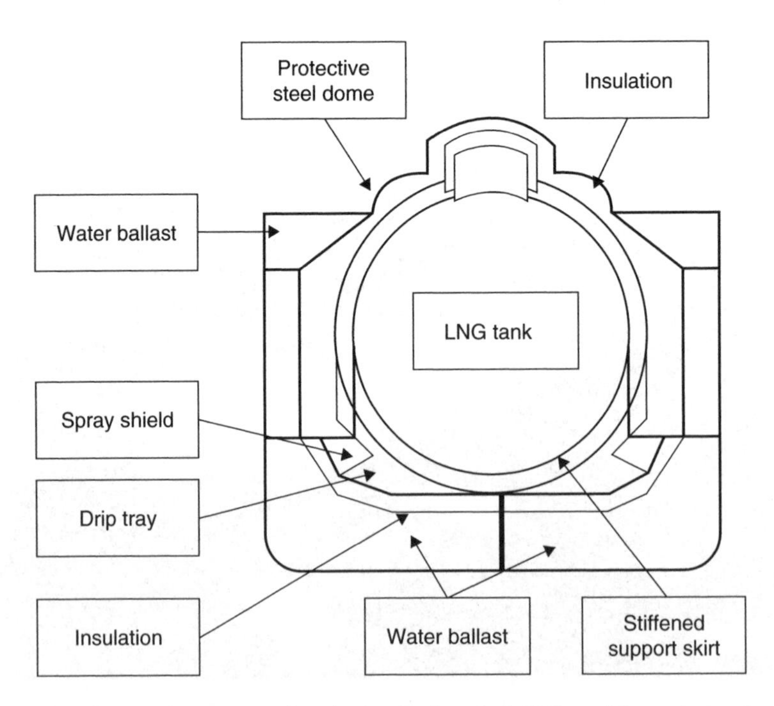

Fig. 5.27 Gas tank construction (spherical tanks). Fully refrigerated spherical LNG tank, the protective steel dome protects the primary barrier above the upper deck (no secondary barrier). Double-hull construction required in way of all cargo tank spaces.

- Integral tanks – tanks which form part of the ships hull
- Membrane – non-self-supporting, completely supported by insulation
- Semi-membrane – non-self-supporting and partly supported by insulation
- Independent tanks – self-supporting tanks not forming part of the ships hull, independent tanks being subdivided into Types A, B and C.

Integral membrane and semi-membrane tanks are designed primarily with plane surfaces. Of the independent tanks, both A and B can either be constructed of plane surfaces or of bodies of revolution, Type C is always constructed of bodies of revolution.

> ***Note:*** *Prismatic tanks (fully refrigerated), carrying cargo at atmospheric pressure, require a primary and secondary barrier to resist undetermined design stresses. The space between the primary and secondary barriers is known as 'hold space' and is filled with IG to prevent a flammable atmosphere in the event of cargo leakage.*

Hazards of gas cargoes

Hazards associated with gas cargoes are from fire, toxicity, corrosivity, reactivity low temperatures and pressure.

Gas carrier types

Gas carrier profile

The more recent builds of LPG carriers include double-hull structure with varied capacity. Up to 100 000 m^3 cargo capacity, is no longer unusual – while LNG construction of 250 000 m^3 using self-supporting, prismatic-shaped tanks requiring less surface space than the normal construction of spherical 'Moss' tanks are under construction with IHI Marine United Shipbuilders (LNG carriage at −162°C and essentially at atmospheric pressure). Cargo boil-off with LNG is used as fuel for the ships propulsion system in some cases or vented to atmosphere (Figure 5.28).

Fig. 5.28 The LPG carrier 'Scott Enterprise' lies port side to a gas storage terminal.

Fully pressurized carriers

These tankers are normally constructed to the maximum gauge pressure at the top of the tank. In all cases, the design vapour pressure should not be less than the maximum, allowable relief valve settings (MARVs) of the tank. This corresponds to the vapour pressure of propane at +45°C, the maximum ambient temperature the vessel is likely to operate in. Relief valves blow cargo vapour to atmosphere above this pressure. Cargo tanks are usually cylindrical pressure vessels. Tanks below deck are constructed with a dome penetrating the deck on which all connections for the loading, discharging, sampling and gauging for monitoring pressure and temperature are placed. Pumps are not normally installed on this type of ship, the cargo being discharged by vapour pressure above the liquid. No vapour reliquefaction facilities are provided.

Semi-refrigerated carriers

This type of vessel is normally designed to carry the full range of LPG and chemical gases in tanks designed for a minimum service temperature of −48°C and working under design pressure. Simultaneous carriage of different cargoes is usually possible. The ships are generally installed with deepwell cargo pumps to facility discharge. If delivery is required into pressurized shore storage units, these deep-well pumps operate in series with booster pumps mounted on deck. Cargo heating using sea water is the usual practice. Vapours produced by heat are drawn off into a reliquefaction unit and the resultant liquid is returned to the tank. This action maintains the tank pressures within limits.

Fully refrigerated carriers

Cargo tanks are usually designed for a minimum service temperature of −50°C and a maximum design pressure.

Discharge of the cargo is achieved by using deepwell pumps or submerged pumps. Unlike the deepwell pumps, the submerged pump assembly, including the motor, is installed in the base of the tank. As a result, it is completely immersed in cargo liquid. Booster pumps and cargo heating may also be installed for discharge into pressurized storage. Reliquefaction plant is also installed on board for handling boil-off vapours. Fully refrigerated carriers now have capacities up to 250 000 m^3.

Fully refrigerated ethylene tankers

The majority of liquid ethylene tankers can carry the basic LPG cargoes as well. Ethylene cargoes are normally carried at essentially atmospheric pressure. Product purity is very important in carriage and care must be taken during cargo operations to avoid impurities, such as oil, oxygen, etc. Reliquefaction plant is also provided on these ships.

Gas operational knowledge

One of the main operational features of working on 'gas carriers' is the awareness of personnel to what is and what is not a gas-dangerous space. This is given by the following definition:

- A *gas-dangerous space, or zone* is a space in the cargo area which is not arranged or equipped in an approved manner to ensure that its atmosphere is at all times maintained in a gas – safe condition.
  - Further: an enclosed space outside the cargo area through which any piping containing liquid or gaseous products passes, or within which such piping terminates, unless approved arrangements are installed to prevent any escape of product vapour into the atmosphere of that space.
  - Also: a cargo containment system and cargo piping.
  - And: a hold space where cargo is carried in a cargo containment system requiring a secondary barrier; a space separated from a hold space described above by a single gas-tight steel boundary; or a cargo pump room and cargo compressor room; or a zone on the open deck, or semi-enclosed space on the open deck, within 3 m of any cargo tank outlet, gas or vapour outlet, cargo pipe flange or cargo valve or of entrances and ventilation openings to cargo pump rooms and cargo compressor rooms.
  - The open deck over the cargo area and 3 m forward and aft of the cargo area on the open deck up to a height of 2.4 m above the weather deck.
  - A zone within 2.4 m of the outer surface of a cargo containment system where such surface is exposed to the weather; an enclosed or semi-enclosed space in which pipes containing products are located. A space which contains gas-detection equipment complying with Regulation 13.6.5 of the IGC Code and space-utilizing boil-off gas as fuel and complying with Chapter 16 are not considered gas-dangerous spaces in this context.
  - A compartment for cargo hoses; or an enclosed or semi-enclosed space having a direct opening into any gas-dangerous space or zone.
- A *Gas-Safe space* is defined by a space other than a gas-dangerous space.

The deepwell cargo pump

Advantages of the deepwell cargo pumps are (Figure 5.29):

1. high speed, high efficiency and high capacity pumps
2. compact in construction when installed in either the vertical or horizontal position
3. choice of power/drive – electric, steam, hydraulic or pneumatic
4. self-flooding
5. automatic self-priming and eliminates stripping problems
6. easy vertical withdrawal for maintenance purposes
7. easy drainage, essential on hazardous cargoes

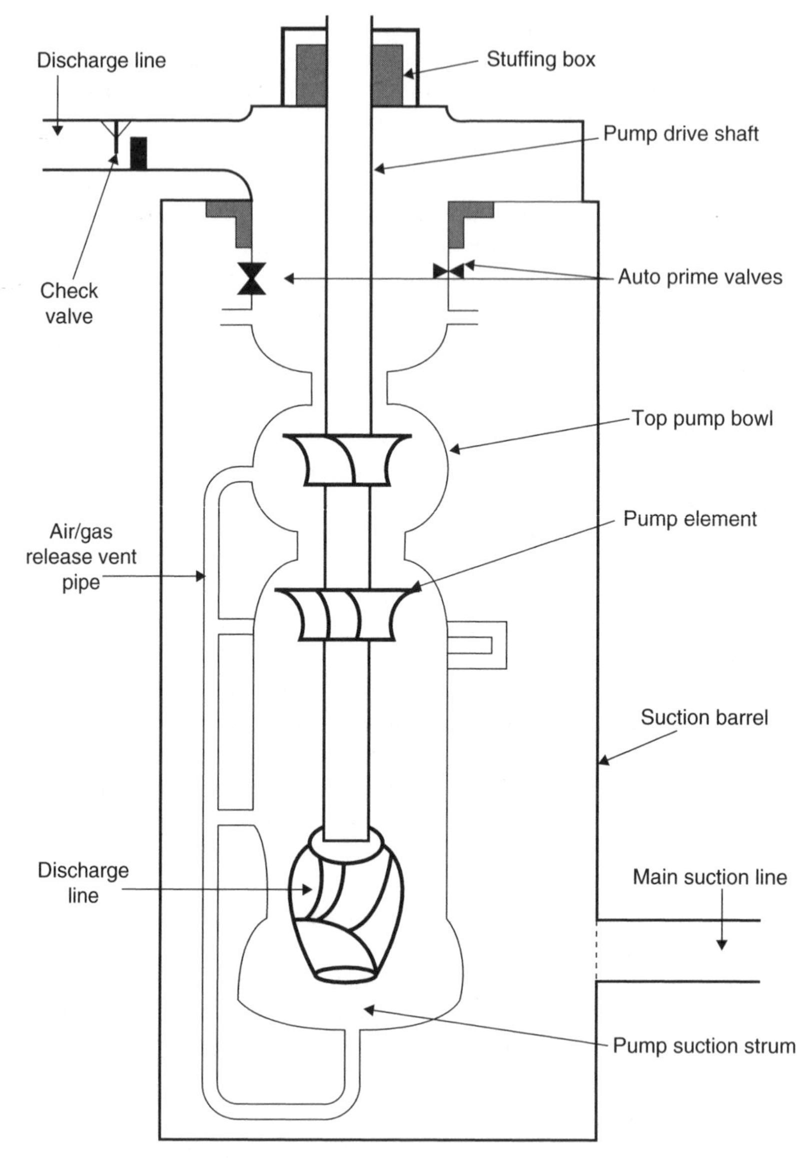

Fig. 5.29 The deepwell cargo pump.

8. tolerance of contaminates in fluid (no filters)
9. improved duty regulation and performance
10. air and vapour locks.

Disadvantages include being suspended, the pump can create construction problems during installation and requires essential rigid bracing supports within the tank, in order to prevent swaying. It also has a long drive shaft which is subject to vibration and torsional stresses.

Deepwell pumps must always be operated and handled in accordance with the recommended operating procedures. The net positive suction head (NPSH) requirements of the pump must always be maintained to prevent cavitation and subsequent pump damage.

LNG carriers

The LNG vessels are normally custom-built for the trade and carriage of the cargo at −162°C and essentially at atmospheric pressure. It is usual for LNG boil-off to be used as fuel for the ship's adopted main propulsion system and they subsequently are not always equipped with reliquefaction plant (Figure 5.30).

Fig. 5.30 LNG carrier. Profile of an LNG vessel seen lying at anchor off Gibraltar harbour. Prismatic tank design as opposed to the spherical tanks.

Cargo operations – safety

Three main safety aspects should be borne in mind when handling liquefied gas:

1. Flammability of the cargo and the need to avoid the formation of explosive mixtures at all times
2. Toxicity of the cargo
3. Low temperature of the cargo which could cause serious damage to the ships hull.

Drying

Once a vessel is ordered to receive a cargo of LNG following overhaul or delivery trials, all traces of water must be removed from the tanks. If this is not done, operating problems due to freezing may result. The dew point of IG or air in equipment must be low enough to prevent condensation of water vapour when in contact with the cold surfaces. Purging with dry gas refrigerated driers and dosing with methanol are not uncommon techniques for removing moisture.

Inerting

Once cargo tanks and associated equipment are suitably dried, air must be removed from the cargo system before loading to prevent the formation of explosive mixtures and also to prevent product contamination. Either IG from the ship's IG generator or a nitrogen supply from shore may be used. IG from a shipboard IG generator is of a relatively low purity content in comparison with 'pure' nitrogen from a shoreside supply and usually will contain up to 15% CO_2 and 0.5% O_2. This can lead to contamination problems with cargoes, such as ammonia, butadiene, etc. To prevent explosive mixture formation, the oxygen content of the tank must be reduced to 6% for hydrocarbon gases and 12% for ammonia using IG or nitrogen.

Purging

When the cargo tanks are suitably inerted, cargo vapours may be introduced to purge the tank of inerts. If the inerts are not completely purged from the tank, then operating problems will be encountered in the reliquefaction plant operations. IG is incondensable and can therefore lead to high pressure in the plant condenser with associated difficulties. The cargo vapours are introduced either at the top or bottom of the tank depending on the density of the gas, and the vapour IG mixture is either vented through the vapour return to the shore flare stack or, where local port regulations allow, to the ship's vent stack.

Cooling of cargo tanks

When about to load liquefied gases into tanks, which are essentially at ambient temperature, it is important to avoid thermal stresses being generated in the ship's structure by incurring high temperature differences. A correct pre-cooling procedure should be adopted to make sure that the tank is brought down in temperature at a rate not exceeding 10°C/h. The most common method of achieving this is to spray cargo liquid from ashore through the tank spray line situated at the top of the tank. This procedure is continued until liquid begins to form on the tank base. Cargo vapours are formed during this cooldown and are either returned through the vapour return line to the shore facility, or, more commonly, handled by the ship's reliquefaction plant on board.

Loading

When tanks have been cooled down, loading of the cargo can commence. Liquid is taken on board via the liquid crossover and fed to each tank through the liquid loading line; this line going to the base of each tank to avoid static electricity build-up. The loading rate is determined by the rate at which the vapours can be handled. Vapours are generated by: (a) flashing of warm liquid; (b) displacement and (c) heat in leak through the tank insulation.

The vapour may be either taken ashore for shoreside reliquefaction or handled by the ship's own plant facility. During the loading operation, cargo tanks must be loaded with regard to trim and stability of the vessel at all times. Cargo tanks must be fitted with high-level alarms to prevent overfilling. Loading rates should be reduced as the cargo levels approach desired values.

Discharging

Discharging can be accomplished by several different methods depending on the equipment which is available aboard the ship:

1. *By use of a compressor alone*: this is usually only associated with small pressurized carriers. The cargo is pressurized from the tank using a compressor taking suction from another cargo tank or with a vapour supply from ashore.
2. *By compressor with booster pump on deck*: the liquid over-pressured from the tank to the suction of the booster pump.
3. *By means of deepwell pumps or submersible pumps*: these are installed in the tank.
4. *By deepwell pumps operating in series with booster pumps mounted on deck*: this is required when discharging into pressurized or semi-pressurized facilities onshore and is carried out in conjunction with a cargo heater for heating the cargo.

An important feature when discharging cargoes is to remember that the cargo is a boiling liquid and will vaporize very easily under normal conditions. When the cargo has been discharged, the vapour remaining in the cargo tank is pumped ashore using the compressor, which would be subject to the design vacuum of the tank.

Working gas cargoes

Certificate of Fitness – an International Certificate of Fitness is required to be carried by any vessel engaged in the carriage of gases in bulk. The certificate is valid for a period not exceeding five (5) years or as specified by the Certifying Authority from the date of the initial survey or the periodical survey (Figure 5.31).

This certificate should be taken to mean that the vessel complies with the provisions of the Section of the Code and is designed and constructed under the International Provisions of 1.1.5 of the Code, and with the requirements of Section 1.5 of the International Bulk Chemical Code.

Fig. 5.31 The Monrovian Gas Tanker 'Annabella' seen starboard side to the Gas Terminal in Barcelona while engaged in gas cargo operations.

The certificate can be issued or endorsed by another government on request. No extension of the 5-year period of validity will be permitted. It will cease to be valid if the ship is transferred to another flag state. A new Certificate of Fitness is issued only when the Government issuing the new certificate is fully satisfied that the ship is in compliance with the requirements of 1.5.3.1 and 1.5.3.2 of the IGC Code.

Cargo conditioning while at sea

During the loaded passage, heat inleak through the tank insulation will cause the cargo pressure and the temperature to rise. The ship's reliquefaction plant should be used to maintain the cargo within the specified limits. In the event that malfunction occurs with the reliquefaction plant, then relief valves will blow vapour off to atmosphere via the ship's vent stack.

Changing cargoes

Depending on the nature of the cargoes it is often necessary to gas free cargo tanks before changing grades.

On completion of the discharge procedure there will always be a little cargo liquid left in the tanks. It is important that this is removed before any gas freeing is attempted, 1 m^3 of liquid forms 300 m^3 of vapour (depending on the substance), and this vapour formation could greatly extend the gas-freeing operation. The residual liquid is blown from the tank by means of an over-pressure created by IG and the liquid is generally blown overside while the ship is on a sea passage.

Fig. 5.32 A loaded LPG carrier, at sea, passing outward bound from the Bosporous and Black Sea Ports. Profile indicates prismatic tanks not spherical tanks.

When the tanks and pipework are known to be liquid free the cargo vapours are swept from the tank using IG or nitrogen.

> ***Note:*** *IG cannot be used when purging because of its high CO_2 content either before loading or after discharging 'ammonia' cargoes. Ammonia would react with CO_2 to form sticky white carbonates.*

The Charter's requirements regarding product purity determine the procedure to be adopted on changing cargoes. Some cargoes, such as VCM, may require visual tank inspection before they can be loaded.

Fig. 5.33 Gas hose sited amidships aboard the LPG carrier 'Scott Enterprise'.

Tank entry

The same general safety requirements relating to tank entry in oil and chemical carriers apply equally to gas carriers.

Reliquefaction plant

The function of reliquefaction plants is to handle vapours produced by heat inflow to the cargo. They are basically a refrigeration plant and may be direct where the vapours are taken through a vapour compression cycle or indirect where the vapours are condensed on refrigerated surfaces, such as cooling coils within or external to the tank (Figure 5.34).

The indirect cycle must be used for gases which cannot be compressed for chemical reasons, e.g. propylene.

The direct cycle can be either single stage or two stages where the cargo condenser is seawater cooled.

Cascade cycle is where the cargo condenser is refrigerated using a suitable refrigerant like Freon 22, within a separate direct expansion cycle.

Pump rooms

To reduce the risk of explosion, cargo compressors and booster pumps are sited in 'pump rooms', divided into at least two compartments with gas-tight bulkheads. The motor's driving compressors are positioned on opposite sides of the bulkheads with the connecting drive shafts fitted with bulkhead seals. Integrity of seals must be monitored and maintained at all times.

> ***Note:*** *Pump rooms are considered as 'enclosed spaces' and as such the full procedures for safe entry into enclosed spaces must be adopted by personnel, as per the CSWP. They are also equipped with emergency escape B/A (EEBA) and full emergency fire-fighting apparatus is readily available.*

Valves

Cargo tanks are protected from over-pressure by relief valves which have sufficient capacity to vent vapours produced under the conditions. Where liquid can be trapped between closed valves on pipework sections, liquid relief valves are fitted to protect against hydraulic pressure developing on expansion.

Liquid and vapour connections on tank domes and crossover are fitted with valves having quick closing actuators for remote operations. These actuators are, in addition, all interlocked with an emergency shutdown system with emergency operational buttons sited throughout the ship.

Instrumentation

Gas carriers are fitted with a gas detector system which continually monitors for cargo leakage. Sampling points are located, for example in void

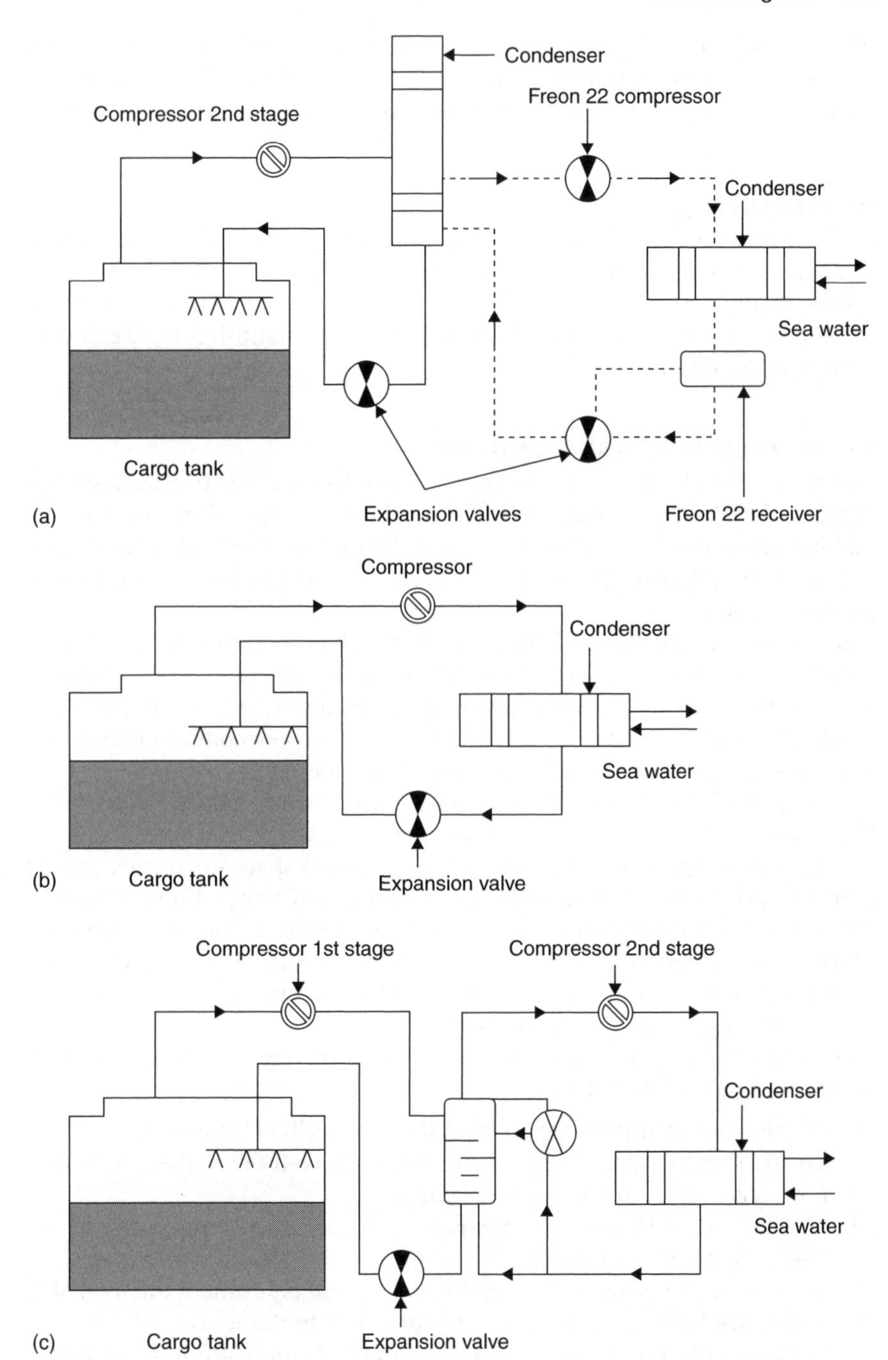

Fig. 5.34 Reliquefaction plant. (a) Direct cycle – cascade system; (b) direct cycle – single system and (c) direct cycle – two stage.

spaces, pump rooms, motor rooms and control rooms, etc. The analyser will alarm on any sampling point reaching 30% of the LEL. In addition, portable gas-detection equipment is provided as describes under oil cargoes.

Fire fighting

Under the IMO Gas Code, gas carriers must be fitted with a water spray system capable of covering such areas as tank domes, manifolds, etc.

Gas carriers must also be fitted with a fixed dry chemical powder system, actuated by IG under pressure having at least two hand-held nozzles connected to the system.

Entry into enclosed spaces

The reader should make additional reference to the Code of Safe Working Practice for Merchant Seaman (MCA publication), regarding the topic of making entry into an enclosed space. A 'Permit to Work' should also be obtained and a risk assessment completed prior to any person entering an enclosed space.

By definition, an enclosed space is one that has been closed or unventilated for some time; any space that may, because of cargo carried, containing harmful gases; any space which may be contaminated by cargo or gases leaking through a bulkhead or pipeline; any store room containing harmful materials; or any space which may be deficient of oxygen.

Examples of the above include chain lockers, pump rooms, void spaces, CO_2 rooms, cofferdams and cargo stowage compartments.

Any person intending to enter such an enclosed space must seek correct authorization from the Ship's Master or Officer-in-Charge. Entry would be permitted in accord with the conditions stipulated by a 'Permit to Work' for entry into enclosed spaces. The Senior Officer would also complete a risk assessment prior to entry taking place and all safety procedures must be monitored by an appropriate safety checklist.

A suggested line of action for permitted entry into enclosed compartments is suggested as follows:

1. Obtain correct authorization from the Ship's Chief Officer.
2. Ensure that the space to be entered has been well-ventilated and tested for oxygen content and/or toxic gases.
3. Check that ventilation arrangements are continued while persons are engaged inside the tank space.
4. Ensure that a rescue system and resuscitation equipment are available and ready for immediate use at the entrance to the space.
5. Persons entering have adequate communication equipment established and tested for contact to a stand-by man outside the enclosed space.
6. A responsible person is designated to stand by outside the space to be in constant attendance while person(s) are engaged inside the space

(function of the stand-by individual is to raise the alarm in the event that difficulties are experienced by those persons entering the space).

7. Ensure that the space to be entered is adequately illuminated prior to entry and that any portable lights are intrinsically safe and of an appropriate type.
8. Regular arrangements for the testing of the atmosphere inside the space should be in place.
9. A copy of the 'Permit to Work' must be displayed at the entrance of the space to be entered.
10. Prior to entry, all operational personnel must have been briefed on withdrawal procedures from the space, in the event that such action is deemed necessary.

> ***Note:*** *When the atmosphere inside an enclosed space is known to be unsafe, entry should not be made into that space.*

Where the atmosphere in the compartment is suspect, the following additional safety precautions should be adopted with the use of 'B/A':

11. Ensure that the wearer of the B/A is fully trained in the use of the B/A.
12. Thorough checks on the B/A equipment must be made and the 'mask seal' on the face of the wearer must be a proper fit.
13. The stand-by man should monitor the times of entry and exit of all personnel to allow adequate time for leaving the enclosed space.
14. Rescue harness and lifeline must be worn.
15. If the low pressure whistle alarm is activated the wearer must leave the space immediately.
16. In the event of communication or ventilation system breakdown, persons should leave the space immediately.
17. Operational personnel should never take the mask of the B/A off when inside the space.
18. The function of the stand-by man is only to raise the alarm if necessary. He should not attempt to affect a single-handed rescue with possible consequences of escalating the incident.
19. Emergency signals and communications should be clarified and understood by all affected parties.
20. A risk assessment must be completed by the Officer-in-Charge to take account of the items covered by the safety checklist, the age and experience of the personnel involved, the prevailing weather conditions, the reliability of equipment in use, the possibility of related overlap of additional working practices ongoing, the technical expertise required to complete the task and the time factor of how long the task is expected to take.

In all cases of enclosed space entry, the use of protective clothing, suitable footwear and the need for eye protection must be considered as an essential element of any risk assessment.

Chapter 6
Specialist cargoes – timber, refrigerated and livestock cargoes

Introduction

The shipping world is actively engaged in trading in virtually every commodity. Many such cargoes fall into specific categories like the container, or the Roll-on, Roll-off (Ro-Ro) trades, and are easy to collate together under a single title or group. However, when attempting to gather all cargoes under one roof so to speak, there is bound to be the odd product that falls outside the norm.

Such cargoes as timber and refrigerated (reefer) and livestock could be discussed to fill a book in their own right. However, the outline of such products falls within the scope of this text which is meant to provide the Cargo Officer with the means to make an educated judgment as to the rights and wrongs of the stowage of these cargo types.

It should be appreciated that cargo-handling methods have changed considerably over the years and the container and Ro-Ro trades have greatly affected quantities of raw products that were previously carried in open stow. No more so than in the refrigerated trades in foodstuffs and perishable goods, many of which are now shipped in refrigerated container units or freezer Ro-Ro trucks.

Timber products in the form of sawn timber in pre-slung bundles or logs can be stored above or below decks. Wood flooring, packaged or pallets may be shipped alongside wood pulp. Such is the variety of timber cargoes. The securing of timber deck cargoes, and the concern for ship security against water absorption is always of concern to a Ship's Master. Timber absorbs great quantities of water at a high deck level, while it burns off tonnes of fuel from low-situated tanks, and could dramatically affecting the ship's metacentric height (GM) and destroy the positive stability of the vessel.

Concern with the specialist cargo must be exercised at all times. It is the duty of the Deck Officer to ensure that not only is the interest of the shipper to be taken account of, but also that of the shipowner, and the well-being of the crew/passengers must be of a high consideration.

Definitions and terminology of specialized cargoes

Absorption – as associated with timber deck cargoes, an allowance made for weight of water absorbed by timber on deck which could have a detrimental affect on the ship's positive stability.

Cant – means a log which is slab-cut; i.e. ripped lengthwise so that the resulting thick pieces have two opposing, parallel flat sides and, in some cases, a third side sawn flat.

CSWP for Ships Carrying Timber Deck Cargoes (IMO 1991) – the Code of Safe Working Practice for the Carriage of Timber Deck Cargoes Aboard Ship.

Freon 12 – is a chlorofluorocarbon (CFC) used as a refrigerant in reefer ships. It is due to be phased out by the Montreal Protocol and is expected to be replaced by a gas (R134a) which has less ozone depletion potential (ODP) and a less greenhouse potential (Freon 22 has already been used in place of Freon 12).

Livestock – a term which describes all types of domestic, farm and wild animals.

Pit props – are straight, short lengths of timber of a cross-section suitable for shoring up the roof in a coal mine.

Reefer – is an expression meant to portray a refrigerated carrier.

Timber – should be taken to mean any sawn wood, or lumber, cants, logs, poles, pulpwood and all other types of timber in loose or packaged forms. The term does not include wood pulp or similar cargo.

Timber deck cargo – means a cargo of timber carried on an uncovered part of a freeboard or superstructure deck. The term does not include wood pulp or similar cargo.

Timber lashings – all lashings and securing components should possess a breaking strength of not less than 133 kN.

Timber loadline – a special loadline assigned to ships complying with certain conditions relating to their construction set out by the International Convention on Loadlines and used when the cargo complies with the stowage and securing conditions of this code.

Wood pulp – and similar substances are not included in the timber terminology as far as deck cargo regulations are concerned.

The air-dried chemical variety must be kept dry, as once it is allowed to get wet it will swell. This action could cause serious damage to the ship's structure and the compartment in which it is carried. To this end, all ventilators and air pipes should be closed off to restrict any possibility of water entering the compartment (stowage factor (SF) 3.06/3.34).

Timber cargoes

Example

Timber is loaded in various forms with differing weights and methods being employed. Package timber is generally handled with rope slings while the heavier logs, depending on size, are slung with wire snotters or chain slings.

Battens – sawn timber more than 10 cm thick and approximately 15–18 cm wide. Usually shipped in standardized bundles and may be pre-slung for ease of handling.

Boards – sawn timber boards of less than 5 cm thick but may be of any width.

Cord – a volume of 128 ft^3 = 3.624 steres.

Deals – sawn timber of not less than 5 cm thick and up to about 25 cm in width. A 'Standard Deal' is a single piece of timber measuring 1.83 m × 0.08 m × 0.28 m.

Fathom – (as a timber measure) equals 216 ft^3 (6 ft × 6 ft × 6 ft).

Logs – large and heavy pieces of timber, hewn or sawn. May also be referred to as 'baulks'. Stowed above and below decks and individual logs may need to be considered as 'heavy lifts' for the safe working load (SWL) of the cargo-handling gear being used.

Pit props – short straight lengths of timber stripped of bark and used for shoring up the ceilings of mines. They are shipped in a variety of sizes.

Stack – a measure of timber equal to half a 'fathom' and equates to 108 ft^3.

Note: *The metric unit of timber measure is known as a 'Stere' and is 1 m^3 or 35.314 ft^3 or 0.2759 cords.*

Timber is generally shipped as logs, pit props or sawn packaged timber. The high SF of timber (1.39 m/tonne), generally indicates that a ship whose holds are full with forestry products will often not be down to her marks. For this reason an additional heavy cargo like ore is often booked alongside the timber cargo. Alternatively, the more common method is to split the timber cargo to positions both below and above decks.

Where timber forms part of the deck stow, some thought should be made to route planning in order to provide a good weather route. Prudent selection of a correct route could avoid prevailing bad weather and unnecessary concerns with the cargo absorbing high seawater quantities. The ship being loaded from the onset with adequate GM and sufficient positive stability could be directly affected in the event of shipping heavy seas in conjunction with timber deck cargoes.

Stowage and lashing of timber deck cargoes

Regulations for the stowage of timber emphasizes that timbe
should be compactly stowed and secured by a series of ovei
adequate strength. Where uprights are envisaged as part o
these uprights should be not more than three (3) metres aŗ
imum height of the timber stow above the uppermost deck must not exceed one-third of the ship's breadth when the vessel is navigating inside a seasonal winter zone.

Additional regulations apply if and when timber loadlines are being used; i.e. when the vessel is being loaded beyond the appropriate normal marks. These regulations take account of timber being stowed solidly in wells at least to the height of the forecastle. If there is no superstructure, at the after end of the vessel, the timber must be stowed to at least the height of the forecastle. This stow must extend to at least the after end of the aftermost hatchway.

A further consideration is that the securing lashings should not be less than 19 mm close link chain (or flexible wire rope of equivalent strength). These lashings shall be independent of each other and spaced not more than three (3) metres apart. Such lashings will be fitted with slip hooks and stretching screws that must be accessible at all times. *Note*: Wire rope lashings must be fitted with a short length of long link chain to permit the length to be adjusted and regulated (Figure 6.1).

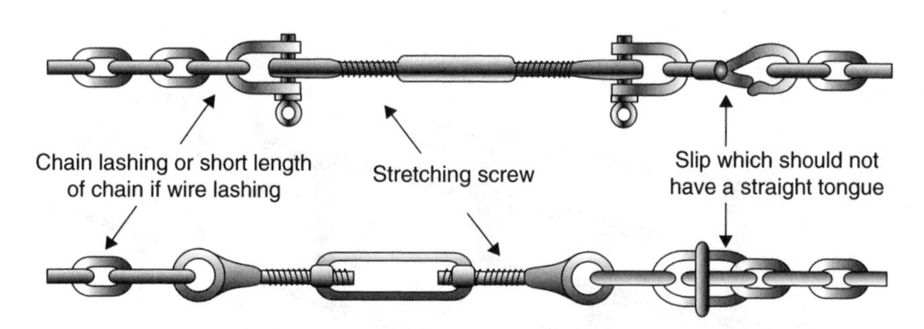

Fig. 6.1 An example of securing lashings.

Additional reference

Additional Reference should be made to Code of Safe Practice for Ships Carrying Timber Deck Cargoes, 1991.

Lashing points

The lashings over timber cargoes are secured to eye plates attached to the sheer strake or deck stringer plate at intervals not exceeding more than 3 m apart. The end securing point to be not more than 2 m from a superstructure ulkhead, but if there is no bulkhead, then eye plates and lashings are to be provided at 0.6 and 1.5 m from the ends of the timber deck stowage position. If the timber is in lengths of less than 3.6 m, the spacings of the lashings are to be reduced. Figures 6.2 and 6.3 indicate some of these points. Access to parts of the vessel fore and aft must be possible and when a capacity deck cargo is carried a walkway over the cargo is generally constructed.

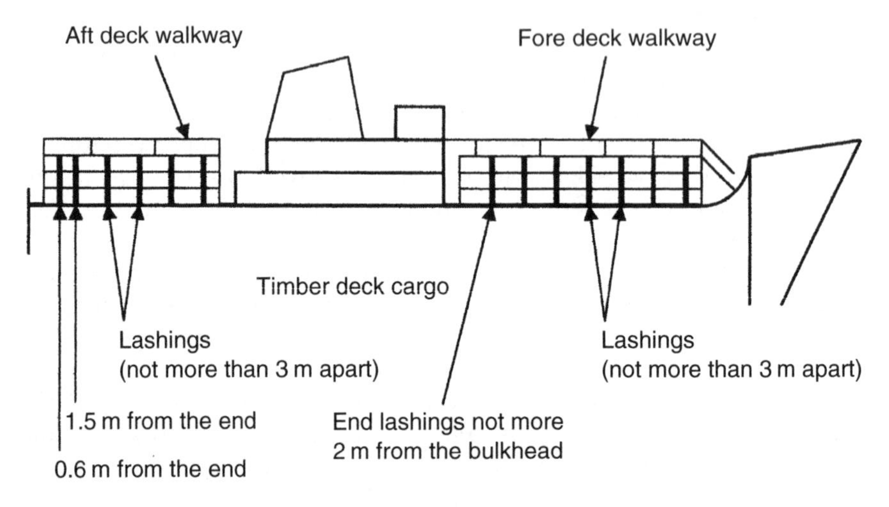

Fig. 6.2 Lashings over timber cargo deck stow.

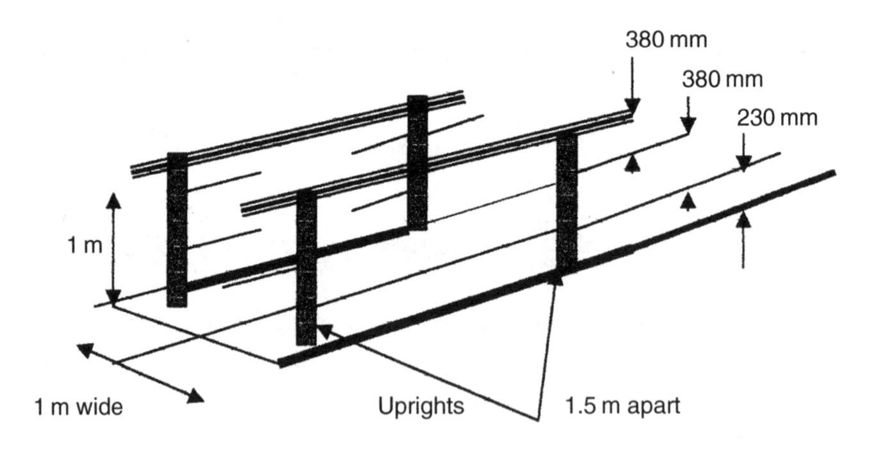

Fig. 6.3 Example walk way construction.

Stowage of logs

The Code of Safe Practice for Ships Carrying Timber Deck Cargoes (Appendix C) provides general guidelines for the underdeck stowage of logs.

Prior to loading logs below decks the compartment should be clean and hold bilges, and lighting tested. A pre-stow loading plan should be prepared considering the length of the compartment and the various lengths of the logs to be loaded.

Recommendations are that logs should be stowed in the fore and aft direction in a compact manner. When loading, they should not be in a swinging motion and any swing should be stopped prior to lowering into the hatch. The heaviest logs should be loaded first and extreme pyramiding should be avoided as much as possible (Figure 6.5).

Fig. 6.4 Packaged timber being loaded onboard ship. Reproduced with kind permission from Everard and Sons, Dartford.

Fig. 6.5 Stowage and the Working of Timber Cargoes. A cargo of logs stowed underdecks in twin hatches, being discharged by multi-fold lifting purchase. Wire snotters are used to manoeuvre the logs to allow slings to be passed under, prior to discharging. Slings are often left in situ after loading to ease discharge but not always. Sawn bundles of lumber are seen stowed at the hatch side as deck cargo.

If void spaces exist at the fore and aft ends of log stows these may be filled with athwartships stowed logs. Logs loaded in between hatch coaming areas should be stowed as compact as possible to maximum capacity of the coaming space.

Logs are heavy and oscillations can expect to cause ship damage. Personnel are advised to maintain a careful watch during the loading/discharging periods.

Packaged timber

Packed timber will usually be banded and may be pre-slung. Packages may not have standard dimensions and may have different lengths within the package, making compact stowage difficult. Uneven packages should not be loaded on deck and are preferred to be loaded below decks. Where deck stowage is made the packages should be stowed in the fore and aft, lengthwise position (Figure 6.4).

Refrigerated cargoes

The increase in container and Ro-Ro trades has, to some extent, brought about the demise of the conventional 'reefer' ship (one that was dedicated to carry refrigerated and chilled cargoes in its main cargo-carrying compartments), the compartments being constructed with insulation to act as very large giant refrigerators. Some of these vessels still operate, particularly in the 'Banana Trade', but generally the cost of handling cargoes into reefer ships has become uneconomic.

Refrigerated cargoes mainly fall into the category of foodstuffs by way of meat, dairy products, fruit, poultry, etc. as a high degree of cleanliness is expected throughout the cargo compartments. Prior to loading such products, the spaces are often surveyed and in virtually every case pre-cooling of dunnage and handling gear has to be carried out. Bilge bays must be cleaned out and sweetened, and the suctions tested to satisfaction. Brine traps should also be cleaned and refilled, brine traps serving a dual purpose by preventing cold air reaching the bilge areas and so freezing any residual water while at the same time preventing odours from the bilges reaching into cargo compartments.

Compartment insulation

All compartments are insulated for the purpose of reducing the load on the refrigeration plant and reducing heat loss from the compartment. It also provides time for engineers to instigate repairs in the event that machinery fails.

Qualities of a good insulation material are that it:

1. should not absorb moisture
2. should not harbour vermin
3. should be fire resistant
4. must be odourless
5. should be low cost and available worldwide
6. should be light for draught considerations
7. should not have excessive settling levels as this would require re-packing
8. should have strength and durability.

Examples in use include: polyurethane, plastics (PVC), aluminium foil, cork granules and glass wool.

Refrigeration plant

Refrigerated cargoes, other than those specifically carried in container or Ro-Ro units, will be carried under the operation of the ship's own refrigeration plant. Cargo Officers are expected to have a working knowledge of the hardware involved with this cooling plant, and the ramifications in the event of machinery failure.

The majority of refrigeration plants in the marine environment operate on the 'vapour compression system' (absorption refrigeration systems are generally not used in the marine environment because they need a horizontal platform).

Figure 6.6 shows a direct expansion, grid-cooling system. A refrigerant like Freon 12 ($C\ CL_2\ F_2$) in its gaseous form is compressed, then liquefied in the condenser. It is then passed through into the grid pipeline of the compartment via the regulator valve. As it passes through the pipes it expands, extracting the heat from the compartment and producing the cooling effect. Its operation is based on the principle that the boiling and condensation points of a liquid depend upon the pressure exerted on it, e.g. the boiling point of carbon dioxide (CO_2) at atmospheric pressure is about − 78°C, by increasing the pressure the temperature at which liquid CO_2 will vaporize is raised accordingly.

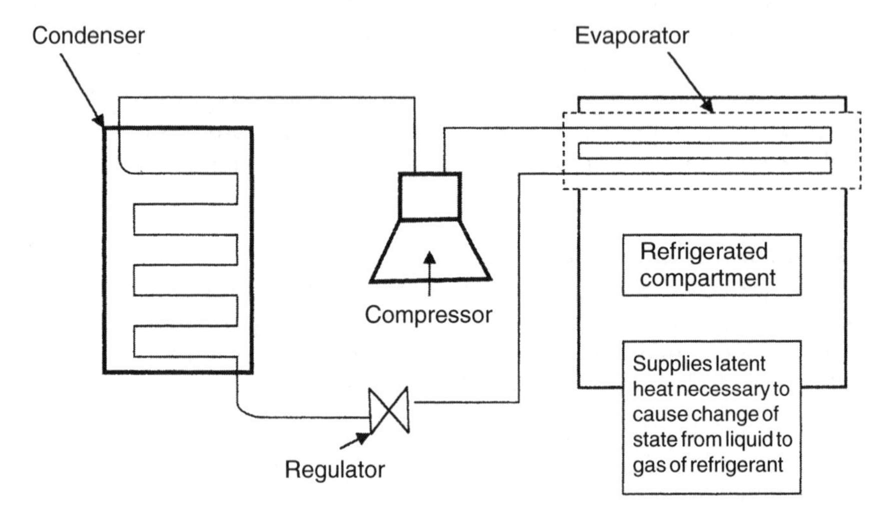

Fig. 6.6 Operation of a vapour compression refrigeration system.

In the past, many refrigerants have been employed in marine refrigeration plants including CO_2, ammonia and more recently the Freon's, but due to depletion of the ozone layer, more refined products are taking over from Freon 12.

Each refrigerant has specific qualities but the popular ones are those having least ODP and less greenhouse potential. It is non-poisonous, non-corrosive and requires only a low working pressure to vaporize and is probably the main one used in any remaining dedicated reefer vessels.

Qualities of a good refrigerant

1. A high thermal dynamic efficiency is required
2. Low cost
3. Low working pressure and low volume
4. Non-toxic, non-inflammable and not explosive
5. Easily available worldwide
6. High critical temperature
7. High value of latent heat
8. Non-corrosive.

Refrigeration plant – monitoring system

In order to protect cargoes, continual monitoring of the refrigeration machinery is considered a necessity. This can be achieved by the introduction of a 'Data Logging System' to the relevant machinery and to the adjoining compartments. With such a system in operation there is less likelihood of damage because an earlier warning system would be activated giving more time to provide corrective action before valuable cargoes are effected by loss of the cooling element.

Sensors and transducers monitor the following points:

1. Temperatures of the cargo compartment
2. Temperature of the fan outlet, discharge air
3. Brine temperatures entering and leaving the evaporator
4. Compressor suction and compression discharge
5. Seawater temperature
6. External air temperature.

Feedback of the sensed parameters are transmitted to either the cargo control room, the engine control station or the navigation bridge (alarm circuits being established to 24 hour manned stations).

Principal refrigerated cargoes and respective carriage temperatures

Product	*Carriage temperature*
Meats: Frozen beef	About −10°C (15°F).
Frozen lamb/or mutton	From about −8° to −10°C (15° to 18°F).
Frozen pork	About −10°C (15°F).
Offal and sundries (includes hearts, kidneys, livers sweetbreads, tails and tongues)	Carried at as low a temperature as possible and not more than − 10°C (15°F). Usually carried in bags or cases. Any of which are blood-stained should be rejected.
Chilled beef	Loaded at about 0–2°C, and carried at about − 1.5°C (29–29.5°F), unless instructed otherwise by the shipper.

Note: Chilling meat only slows the decomposition process down and it remains in prime condition for about 30 days. This period could be extended by about 15 days if a 10% concentration of CO_2 is introduced into the compartment, assuming the compartment can be sealed and the environment is safe to permit such action.

Poultry	Packed in cases and carried at −10°C to −12°C (10–15°F).
Dairy products	
Butter	Liable to taint and should not be stowed alongside other strong smelling cargoes in the

	same compartment, e.g. fruit. Generally packed in cartons. Carriage temperature about −10°C (15°F).
Cheese	Carriage temperature varies but generally carried at 5–7°C average. Usually stowed on double dunnage.
Shell eggs	Stored in cases and liable to taint. Normally not stowed above 10 cases high with air circulation channels on top of 50-mm dunnage. Carriage temperature 1°C (33°F).
Liquid eggs	Carried in tins at temperatures not over −10°C (15°F).
Bacon	Stow on double 50-mm dunnage, do not overstow. Carrying temperature −10°C (15°F).
Fish	Shipped in boxes or crates and should be stowed on 50-mm dunnage. Fish has a tendency to rapid deterioration, and should be carried at a low a temperature as possible, which should not exceed −12°C (10°F).

Fruits

Fresh fruits are generally carried in cardboard cartons or wood boxes, with ventilation holes. They can often be carried in non-refrigerated spaces on short haul runs Good ventilation must generally be given to prevent a concentration of CO_2 build-up. CO_2 must not be allowed to build up over 3% concentration as this would cause deterioration of the cargo. Frequent air changes are recommended to avoid this.

Apples	Carriage temperature will vary with the variety of apple but is usually in the range of −1–2°C.
Pears	Should not be stowed in the same compartment as apples. Carriage temperature −1°C to 0°C (30–32°F).
Grapes, peaches, plums	Carriage temperature −1°C to 2°C (31–35°F).
Oranges	Oranges must have adequate ventilation as they are very strong smelling and the compartment must be deodorized after carriage. Carrying temperature 2–5°C (36–41°F).
Lemons	Similar to oranges. Carrying temperature 5–7°C (41–45°F).
Grapefruits	Similar stow to oranges. Carriage at about 6°C (44°F).

Bananas	The banana trade is specialized and special ships are built for the purpose. Many of which use containers.The carriage temperature is critical as too low a temperature can permanently arrest the ripening process. Daily inspection of a compartment would be carried out and any fruit found to be ripe is removed. One ripe banana in a compartment can cause an acceleration of the ripening process throughout the compartment. Carriage temperature usually about 12°C (52–54°F).

The 'reefer' trade

It should be realized that many of the said cargoes are now shipped by refrigerated containers or Ro-Ro cold units. Some companies still operate designated refrigeration vessels like those employed by 'Lauritzen Cool' engaged on the New Zealand to the US West Coast meat service.

Other specialized parcels, like some drugs, often require refrigerated stowage and the instructions as to the carriage temperature would be issued by the shipper.

Prior to loading any refrigerated cargoes it is normal practice for a surveyor to inspect the compartment for cleanliness and to ensure that the compartment temperatures are correct. Dunnage and any cargo fitments would be pre-cooled and machinery would be tested to satisfaction. Cooled gas and chemical cargoes are referred to in Chapter 5.

Refrigerated container units

Lloyds Register have developed 'Rules for the Carriage of Refrigerated Containers in Holds'. These standards take account of the problem of heat emanating from an on-line refrigeration plant operating below decks in the cargo hold. The heat energy rejected by each unit is from the evaporator fans, the motor and the condenser. Concern for this rejected heat energy into the surrounding air of the hold is currently considered a problem that may or may not be resolved by improved ventilation methods.

The container sector of the industry is exploring ways to carry increased numbers of reefer units below decks. However, such increase would generate increased temperatures into the cargo space areas. An effective ventilation system would probably aim to retain the hold temperature as close as possible to the outside air temperature or a predetermined temperature to suit the internal hold environment.

The majority of refrigerated containers employ insulation, usually polyurethane, within the prefabricated construction of the container. This directly affects the heat transmitted through the insulated unit between the carriage temperature and the external ambient air temperature. Although the insulation will reduce the actual payload capacity of the unit, it is seen as a necessary trade off.

Example

Carriage temperatures for a 40-ft container:

Bananas	13.0°C
Chilled apples	2.0°C
Frozen	−18.0°C
Deep frozen	−29.0°C

Various ventilation systems operate throughout the industry. The one illustrated in Figure 6.7 is a semi-sealed louvred exhaust duct system. A vertical ducting fitted with an air supply fan delivers supply air to each stack of containers, specifically to each container condenser. The exhaust system operates in a similar manner, with the exception that the fan is an extraction fan as opposed to a supply fan. Isolation valves or flaps could be fitted to isolate 'cells' when not in use, each cell having its own inlet and outlet ducting (at time of writing modelled format only).

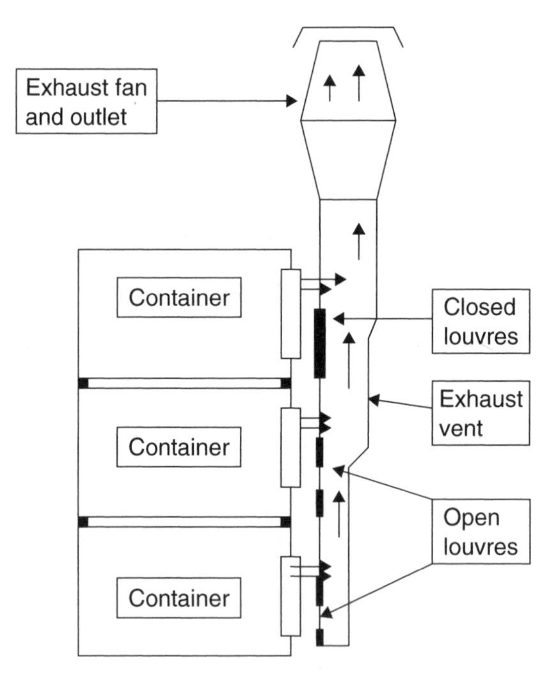

Fig. 6.7 Refrigerated Container – Hold Ventilation System.

The carriage of livestock

The carriage of animals, either domestic, farm or from the wild, is not an uncommon practice. The carriage is governed by the regulations laid down by the Ministry of Agriculture and Fisheries. Further advice can also be obtained from various animal protection societies who give advice on cage size, crates, etc. for use with animals.

Where large numbers of animals are to be carried – like sheep or cattle – designated livestock carriers are available. The ships tend to discharge the beasts directly into penned, quarantine areas. While in transit the animals are kept in pens or stalls which are protected from adverse weather and the sun.

Adequate straw and fodder would also be carried. The feeding and watering of animals would be to the shipper's instructions. It is not unusual for a shipper to send a supervisory attendant where large numbers of animals are carried or where specialist animals like valuable race horses are carried. If no attendant is carried, members of the crew would be designated to take care of the animals during the passage, cleaning stalls and feeding, etc.

Where one or two animals are carried by a non-designated vessel, they are usually carried in horse box-type stalls, or in caged kennels. These are generally kept on a sheltered area of the upper deck away from the prevailing weather. Each animal would be tallied and allotted a carriage number. In the event of the animal dying on passage, this number must be recorded. All vessels carrying livestock must carry a 'humane killer' with enough ammunition to be considered adequate.

Where a regular livestock trade is featured, like Australia/Middle East regions, shore facilities for loading and discharging are regularly inspected by the country's authorities. Ministry officials also inspect the cleanliness and the facilities aboard designated livestock carriers.

Documentation inclusive of veterinary certificates is usually shipped with the animal(s) together with routine welfare instructions. When landed, documentation is usually landed at the same time being handed to the shipper's representative or quarantine officials.

Chapter 7
Roll-on, Roll-off operations

Introduction

Some time after the start of containerization came a cargo revolution in the door-to-door service of Roll-on, Roll-off (Ro-Ro) handling procedures. The Ro-Ro traffic provided a shuttle service for containers as well as cutting delivery times to hours rather than weeks, previously experienced with conventional shipping. The Ro-Ro explosion was so great that ports changed their operations and ship design started to incorporate new concepts, to handle large vehicles.

The coastal traffic saw a new lease in life which opened up numerous avenues, in employment, cargo-handling methods, service industries and manufacturing. Ferry companies increased their tonnage maximums in a comparative blink of an eye. Port exports climbed beyond previous records, with Ro-Ro activity being the main cause. Ro-Ro was an efficient and cheap method of shipping merchandise which was quickly realized and expanded rapidly beyond anyone's wildest expectations.

The ship's new design included the stern door/ramp, open vehicle deck spaces, drive through capability with the bow visor. Vehicle lifts became a feature with open and enclosed deck cargo spaces. Units could be carrying liquid or dry cargoes, they could be refrigerated or not, as their load required. However, the most important fact was that they could be delivered in the shortest period of time.

The time factor was critical to ensure that goods reached markets in a pristine condition. Especially relevant to fresh produce like flowers, fruits, dairy foods, meats, etc. The ships were enhanced to ensure that deadlines were achieved. Ships docking in and out carrying such cargoes could not be delayed by the need for tugs. Bow/stern thrusters became essential features of ship design. Thruster units came alongside twin Controllable Pitch Propellers, while Masters were given Pilotage Exemption Certificates. Not only were the vessels fast, but also the procedures and concepts of ship handling had been changed to meet the needs of the trade.

The Ro-Ro trade has now become an essential segment of the shipping industry. Although it might be seen as the new boy on the block, it is already alongside the tanker traffic, the cruise trade and greatly attached to

Fig. 7.1 Ro-Ro shipping. A typical Ro-Ro ferry the 'Stena Leader' (previously European Leader and ex-Buffalo, P&O Ferries) departs the Port of Fleetwood for her regular twice-daily voyage to Ireland. The vessel carries about 140 mobile units (40-ft container size) on three vehicle decks. The ship also accommodates a limited number of unit drivers.

the container business. The sector ships everything that was once shipped by general cargo vessels. These cargoes include hazardous goods, as well as heavy-lift units. The main difference is that such items are controlled by separate legislation and generally move with less bureaucracy (Figure 7.1).

Ro-Ro definitions and terminology

Freight only Ro-Ro ship – a Ro-Ro vessel with accommodation for not more than 12 (driver) passengers.

High-speed craft – a craft capable of a maximum speed, in metres per second (m/s). Equal to or exceeding 3.7V × 0.1667 where V = displacement corresponding to the design waterline (m^3).

Passenger car ferry – a passenger or ferry ship which has Ro-Ro access of sufficient dimensions to allow the carriage of Ro-Ro Trailers and/or Ro-Ro Passenger (Ro-Pax)/Ro-Ro Cars (Figure 7.2).

Reefer unit – a mobile/vehicle Ro-Ro unit, designed and capable of carrying refrigerated cargoes.

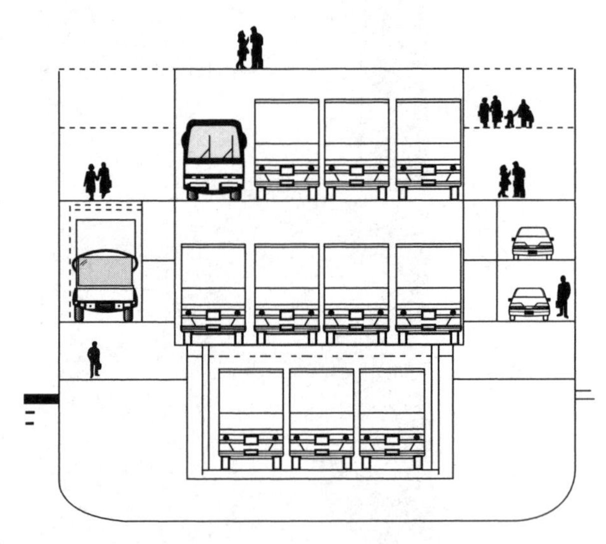

Fig. 7.2 Diagram of passenger car ferry.

Right of ferry – an exclusive right to convey persons or goods (or both) across a river or arm of the sea and to charge reasonable tolls for the service.

Ro-Ro cargo space – a space not normally subdivided in any way and extending to either a substantial length or the entire length of the vessel in which goods are carried (packaged or in bulk), in or on rail or road cars, vehicles (including road or rail tankers), trailers, containers, pallets demountable tanks in or on similar stowage units or other receptacles, can be loaded and unloaded normally in a horizontal direction.

liverpool
JMU

LJMU

Roll-on Roll-off vessel – a vessel which is provided with horizontal means of access and discharge for wheeled, tracked or mobile cargo (Figure 7.3).

Fig. 7.3 The modern face of Ro-Pax type vessels. High-speed catamaran or tri-maran hulls with vehicle access from a stern ramp. Generally engaged on the short sea trades around the world, operating at service speeds up to 45 knots.

Short international voyage – an international voyage in the course of which a ship is not more than 200 nautical miles from the port or place in which passengers and crew could be placed in safety. Neither the distance between the last port of call in the country in which the voyage begins and the final port of destination, nor the return voyage, shall exceed 600 nautical miles. The final port of destination is the last port of call in the scheduled voyage at which the ship commences its return voyage to the country at which the voyage began.

Special category space – any enclosed space, above or below the bulkhead deck intended for the carriage of motor vehicles with fuel in their tanks for their own propulsion, into and from which such vehicles can be driven and to which passengers have access (Figures 7.4–7.7).

Vehicle ramps

The design of Ro-Ro vessels is influenced from the onset of the design stage by the nature of the payload it is intended to transport. Generally, the cargo flow, securing and handling equipment can amount to about 5% of the lightweight

Fig. 7.4 Modern Ro-Ro (freight only) vessel. The 'MYKONOΣ' a modern Greek operated Ro-Ro vehicle ferry. Design features include all accommodation forward, with twin Port and Starboard smoke stacks seen aft either side of the upper vehicle deck. The stern door/combined vehicle ramp is positioned right aft in the upright closed position while the vessel is at sea.

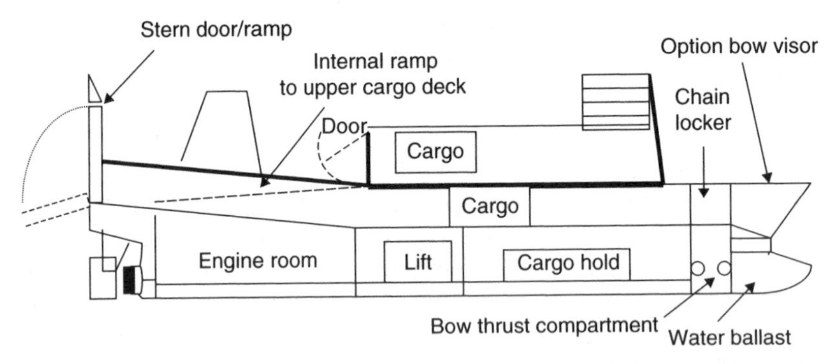

- Open deck stowage of Ro-Ro cargo either side of the engine room smoke stack.
- Chain locker at the ships sides either side of the fore and aft line to facilitate the operation of separate Port and Starboard windlass operations.
- Bow visor option to permit drive through capability and is not always featured.
- Lift to lower cargo hold may be mechanical or of hydraulic operation.
- All cargo ramps are fitted with wheel tread, anti-skid, steel grips.
- All cargo decks are fitted with insert star lashing points and/or star domes.
- Accommodation for twelve (12) driver/passengers.

Fig. 7.5 General arrangement – modern Ro-Ro ferry (freight only) 1900-m lane length.

Fig. 7.6 Ro-Ro ferry example. Stern door/ramp access into the enclosed vehicle deck of a Ro-Pax ferry operating in the Mediterranean Sea.

Fig. 7.7 The bow visor of the passenger vessel 'Jupiter' (since renamed) seen in the open position against the skyline. The bow visor fitted with a stern ramp access permits a drive through capability.

tonnage. However, to avoid operational problems in the future such fittings need to take account of the types of rolling cargo which is anticipated. Commercial vehicles are limited to about six types (unlike military vehicles) and these need to be accommodated by respective access widths, ramp slopes, clearing heights, lane lengths, turning areas or drive through facilities.

Similarly, shoreside receptions must be compatible with a ship's facilities. Ramp slopes and break angles, for commercial traffic, will generally fall at about 1 : 8 or 1 : 10 in order to avoid the vehicle grounding while in transit from the ship to the shore. Where tidal waters are present and average rise or fall is expected, floating shore links or adjustable link spans tend to overcome excessive tidal movement, while at the same time keeping the break angle with the ship's ramp manageable (Figure 7.8).

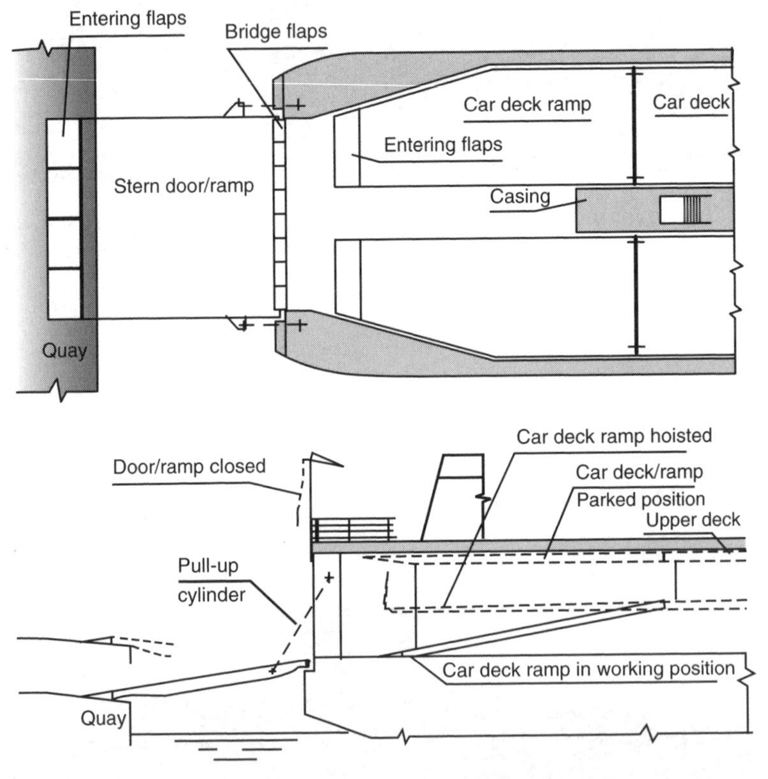

Fig. 7.8 Example stern door and vehicle ramp arrangement.

The design of equipment will be to the requirements of Lloyds Register or similar Classification Society, but would include specific features to satisfy operational needs. In order to match these needs and provide a suitable end product, a designer would include the following features:

1. Length of ramp (overall)
2. Width of ramp (overall)

3. Total load on ramp (anticipated maximum)
4. Maximum axle loads
5. Hinging arrangement (top, bottom or guillotine)
6. Number of ramp sections and hinges within the structure
7. Maximum/minimum operating angles
8. Watertight sealing/securing arrangements
9. Cleating/locking arrangements
10. Power requirements (electric, hydraulic) with limitations
11. Operational lifting/lowering times
12. Supporting and preventor arrangements
13. Roadway landing area.

Many stern ramp arrangements open up all the transom to provide maximum width and height clearance. This effectively gives wide access to a variety of vehicles of differing lengths with comparative short load/discharge times involved. Other designs have employed stern quarter ramps (with or without bow quarter ramps). Such ramps are still required to meet the design criteria of the Classification Society but must also satisfy design features to meet specific vehicle traffic like 'car carriers' (Figure 7.9).

Ramps tend to be manufactured in steel with 'Chevron Pattern' anti-skid bars on the working surface. They are usually operated by twin hydraulic

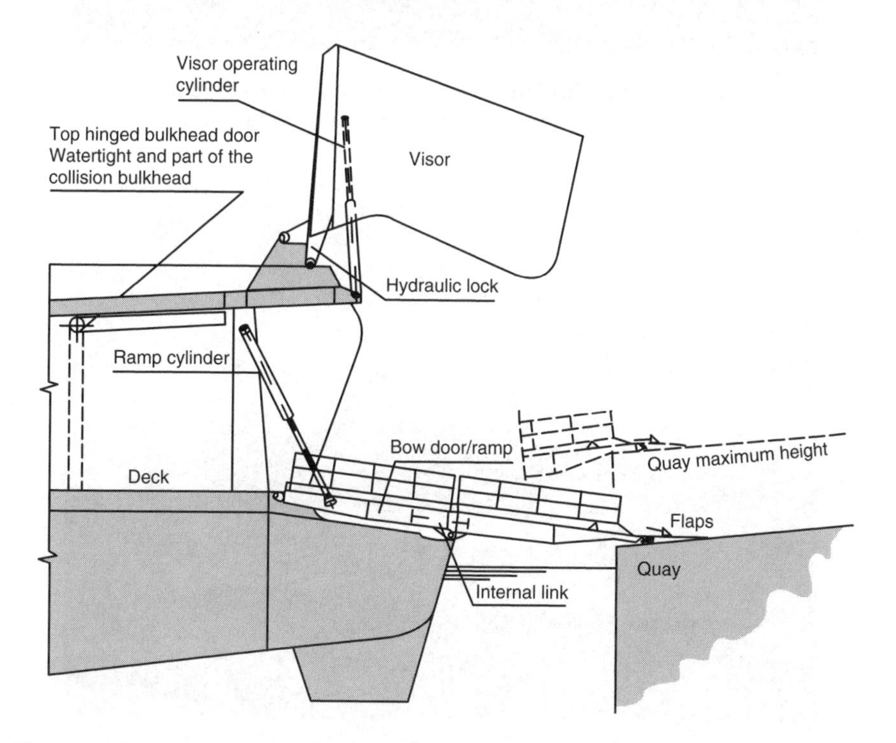

Fig. 7.9 Bow visor with combination inner bow door and vehicle ramp arrangement.

cylinder actions or winch arrangement. Watertight integrity is achieved with hydraulic pressure cleating in conjunction with a hard rubber seal, with the hinge arrangement being positioned above the waterline (Figures 7.10 and 7.11).

Fig. 7.10 Stern door of a Ro-Pax ferry seen in the stowed, closed position as she turns off the berth in the harbour at Tangier.

Fig. 7.11 The Ro-Pax ferry 'Sanasa' of the Comarit Ferry Group enters Tangier harbour. The vessel is fitted with a bow visor and stern/ramp door.

Fig. 7.12 An internal ramp that can tilt both forward and aft to suit stern load and/or bow discharge. It allows loading to a higher deck level 'three' where the lower hold level would be termed 'No. 1 Deck Level' and 'No. 2 Deck Level' would be considered as the main, largest of the three vehicle decks.

Internal ramps and elevators

The current generation of Ro-Ro vessels have moved into multi-deck construction with a totally enclosed main vehicle deck with access from either a stern ramp or bow door arrangement. This deck is often fitted with elevator access to a lower hold while an internal ramp to a partially covered upper vehicle deck, which permits access to the higher, uppermost continuous deck (Figures 7.13 and 7.14).

Fig. 7.13 Cargo doors. Upper deck hydraulic cargo door set in the bridge front and leading to an enclosed upper vehicle deck aboard a modern Ro-Pax ferry. Weather sealing of the door takes place against hard rubber seals with hydraulic cleating and positive pressure held on operational rams. When in the open position the door is locked with hydraulic securings and rams are fitted with non-return valves to prevent accidental closing.

Fig. 7.14 Example vehicle deck of the enclosed cargo compartment of a high-speed passenger vehicle ferry operating in the Irish Sea region of the UK coastline. Steel pillars support aluminium upper decks and the internal, angled ramp is seen at the upper right-hand side of the view. The deck structures are manufactured in unpainted aluminium to save increased weight.

Internal cargo operations – Ro-Ro vessels

Vehicles require wide open deck space to be able to manoeuvre. Such deck areas are lane marked to ease vehicle stowage and alignment of mixed types of vehicles, e.g. private cars and commercial trucks. The deck areas are always well-illuminated by overhead lighting and fitted with extraction fans to change the air volume 10 times every hour. Such atmosphere replenishment prevents the build of exhaust gases from drive-on, drive-off operations.

Cargo vehicle decks are protected by sprinkler and/or water-drenching systems and well provided for with fire extinguishers at every 40 m length. Such protection dictates that the decks must also be fitted with an adequate drainage system to clear residual waters quickly (Figure 7.15).

Vessels without 'bow visor' facilities are generally denied drive through capabilities and usually must provide sufficient deck space to permit the turning of wagons ready for stern discharge at the arrival port. Vehicle decks have always been considered as a hazardous environment for both shore and shipboard personnel, especially where vehicles are turning. To this end speed of vehicles is strictly controlled by stowage marshals who usher units into designated lane spaces. With this in mind deck spaces are clearly sign painted to reflect basic instructions to driver personnel and car/passenger travellers (Figure 7.15).

Fig. 7.15 Typical bulkhead markings prominently displayed around vehicle decks to ensure safe and efficient loading of vehicle lanes.

Fig. 7.16 Ro-Ro traffic trailer units, parked on the dock side at Liverpool ready for loading on to P&O, Irish sea ferries. Trailer units being lifted by mechanical 'tugs' and deposited on the vehicle decks of short sea ferries then similar 'tugs' attach to discharge the unit at the destination port.

Ventilation system

It is a requirement that Ro-Pax vessels carrying more than 36 passengers must be provided with a powered ventilation system (fans) sufficient to give 10 air changes per hour in spaces designated to carry vehicles (with fuel in their tanks for their own propulsion). If the vessel carries less than 36 passengers then the venting system need only provide six air changes per hour.

Ventilation ducting serving such spaces should be constructed in steel, and the system should be completely separate from other ventilation systems aboard the vessel. It must be capable of being controlled from outside the vehicle spaces and be operable at all times when vehicles are occupying the specific areas.

> ***Note:*** *Where special category spaces are employed, the administration may require an additional number of air changes when vehicles are being loaded or discharged.*

Ventilation systems must be fitted with rapid means of shut down, in the event of fire occurring. They must also have a means of monitoring any loss or reduction in the venting capacity with such data being indicated on the 'navigation bridge'.

Drainage systems

The hazards of slack water on large vehicle decks and the subsequent loss of stability which could occur are well known. The fixed pressure water spraying system, installed for fire prevention, if operated, could cause an accumulation of water on vehicle deck or decks. To ensure adequate stability at all times a suitable drainage system must be installed to effect rapid discharge of slack water.

Scuppers should be fitted to ensure discharge directly overboard. Special category spaces situated above the bulkhead deck, and in all Ro-Pax vessels which have positive means of closing scuppers by valve action, must keep such valves open while the vessel is at sea in accord with the 'loadline convention'.

In the case of special category spaces, the administration may require additional bilge pumping and drainage facilities over and above the specifications of Safety of Life at Sea (SOLAS), Regulation II-1/21.

Bilge pumping arrangements

Cargo ships and passenger vessels are required to have in place an efficient bilge pumping system, capable of pumping from and draining any watertight compartment. Passenger Ships are required to have at least three (3) power pumps connected to the bilge main.

Cargo (vehicle) definitions

A vehicle – defined as a vehicle with wheels or a track laying vehicle.

A flat-bed trailer – defined as a flat-topped open-sided trailer or semi-trailer and includes a roll trailer and a draw-bar trailer.

Freight vehicle – defined as a vehicle which is a goods vehicle (flat-bed trailer) (road train) (articulated road train) combination of freight vehicles or a tank vehicle.

A semi-trailer – defined as a trailer which is designed to be coupled to a semi-trailer towing vehicle and to impose a substantial part of its total weight on the towing vehicle.

A tank vehicle – defined as a vehicle fitted with a tank which is rigidly and permanently attached to the vehicle during all normal operations of loading, discharging and transport and is neither filled nor discharged on board and driven on board by its own wheels.

Reefer unit – container box unit fitted with refrigeration plant. Employed to transport frozen/chilled produce by road and sea. Power for the freezer unit is generated by the drive motor of the unit when on the road and supplied from the ship's supply while the vessel is at sea. (Special stowage space is required for reefer units to ensure that they are positioned aboard the ferry close to a power supply connection.)

Fig. 7.17 Long load (sail for wind turbine) is loaded on adjustable stretch load trailer unit.

Ro-Ro vehicle types

The majority of freight vehicles engaged in Ro-Ro vessels vary in size and type for the shipment of cargoes, both dry goods and liquids. Probably the most widely used is the drop trailer vans (40-ft container box/van stowed on a horse or trestle and the rear wheels of the unit). Other varieties include:

Curtain-sided trailers
Semi-trailer without sideboards (drop sides)
Semi-trailer with sideboards
Semi-trailer with sideboards and hood cover
Fully enclosed goods vehicle
Open flat-top truck
Flat-top truck with canvas-covered load
Articulated trailer
Road tanker
Framed container/tank
Freight container (20 × 8 × 8)
Freight container (400 × 8 × 8)
Draw-bar combination (two units)
Draw-bar combination (three units)
Refrigerated (reefer) vans
Low loaders (for heavy machinery/plant)
Adjustable (stretch) low loader (for exceptional long loads) (Figure 7.17).

Additional private vehicles such as coaches, furniture removal vans, buses, caravans, boats on trailers, military transports, etc. are also regularly shipped. Freight units of one kind or another, once discharged, may be reloaded but not in every case. Many units are often returned by the same ferry or a sister vessel in an empty state (Figures 7.18 and 7.19).

Fig. 7.18 Ro-Ro, unit types. Several Ro-Ro units on the quayside in Cadiz, Spain. From left to right: a flat-top trailer unit, a 40-ft container carrier, with two articulated container carriers on the end. A mobile tank container trailer unit is seen in the background.

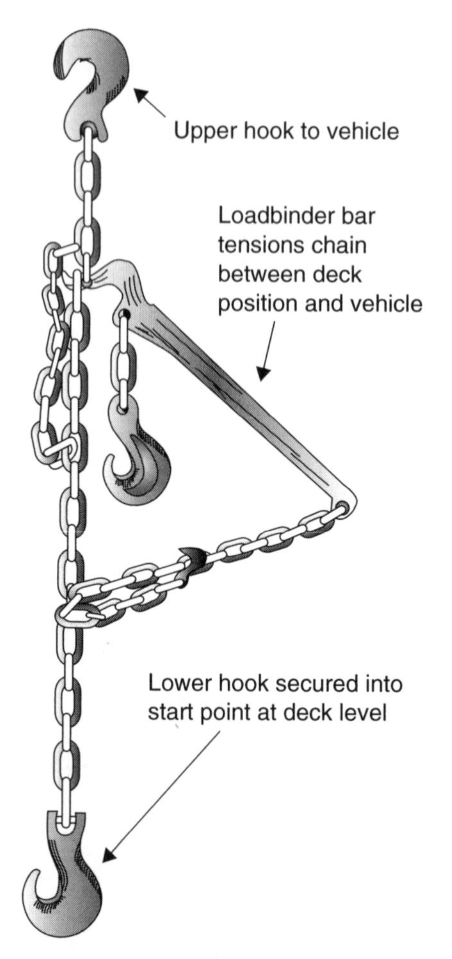

Fig. 7.19 Example of a chain lashing.

Vehicle stowage and securing

It is essential with vessels-carrying vehicles that a stable deck is maintained and this is why virtually all Ro-Ro ferries are now built with stabilizers of one form or another. However, cargo movement can still expect to occur in very rough sea conditions even when stabilization systems are operational. To this end individual vehicles are secured by various means to prevent movement at sea.

The stowage/securing arrangements of units should be supervised by a responsible Ship's Officer assisted by at least one other competent person. Vehicles should, as far as possible, be aligned fore and aft, with sufficient distance between vehicles so as to allow access through the vehicle deck. The parking brake on each vehicle/unit should be applied and where possible the unit should be placed in 'gear'. Where drop loads or uncoupled units are being carried these should be landed on trestles or equivalent support, prior to being secured by chain or other suitable securing constraint (Figure 7.21).

All vehicle/cargo units should be secured prior to the vessel leaving the berth and such securings should be at the master's discretion to be most effective. While on route these lashings should be regularly inspected to ensure they remain effective during the time at sea. It should also be realized that personnel so engaged on vehicle deck inspections should take extreme caution against injury from swaying vehicles. As such, Masters may feel it appropriate to alter the ship's course while such inspections are ongoing to reduce the motion on the vehicle deck.

Vehicles stowed on slanting decks should have the wheels 'chocked' and the hand brakes observed to be on and working. Suitable lashings against the incline should be secured and the unit left in an opposing gear. Any vehicle which is lashed should be secured at the correct securing points so designed on the vehicle and at the deck position.

All lashings applied whether of a 'hook' type or other variety should be secured in such a manner that in the event of them becoming slack, they are prevented from becoming detached. They should also be of a type which will permit tensioning in the event of them becoming slack during the voyage.

Note: *Lashings are considered to be most effective at between 30° and 60° to the deck line. Alternatively, additional lashings may be required. Crossed lashing should, where practical, not be used, as limited restraint against 'tipping' is experienced with this style of securing.*

Lashings should only be released once the ship is secured at the berth and personnel so engaged should take care when clearing securings. These may be under high tension following transit and cause injury if released without forethought.

Note: *Cargo units must be loaded, stowed and secured in accord with the Ship's* Cargo Securing Manual, *as approved by the Authority. (This Cargo Securing Manual is required to be carried aboard all types of ships engaged in the carriage of all cargoes, with the exception of bulk cargoes.)*

Unit securing – chain lashings

Ro-Ro units are secured in accordance with the Cargo Securing Manual of the vessel. In some short sea voyages, during the summer season and with a predominantly good weather forecast, units may not even be secured other than by the hand brakes and left in gear. However, at the Master's discretion, chain lashings could be applied by the crew if and when circumstances dictate that securing becomes necessary.

In virtually all cases, hazardous units would automatically be chained down. Chain lashings vary but tend to have a common theme of being able to be applied between a deck 'star' lashing point and the unit itself, then tensioned by a load-binding lever.

Such lashings can be secured and tensioned quickly, and lend to labour saving. The number of lashings per unit will be variable, depending on the weight and size of the vehicle. However, a standard 40-ft unit would usually be fitted with a minimum of six (6) lashings.

Vehicle decks are built with star lashing points or 'elephants feet' type anchor points. Lashings will have a club-foot fitting into these points, with a hook at the opposite end. Alternatively, as shown in Figure 7.19, hooks at each end.

Ro-Ro ship stability

Modern Ro-Ro shipping has experienced some painful losses over the years, the most notable being the Herald of Free Enterprise (1987) the Estonia (1994) and more recently the Tricolour (2003) with 2800 cars, and the Hyundia No. 105 (2004) with more than 4000 vehicles on board. Clearly, the losses and subsequent salvage operations have rocked the marine insurance markets generating tighter legislation to cause improved conditions on Ro-Ro vessels.

Improvement features now include the following:

1. The stability of the vessel must be assured as adequate, with the main deck flooded to a depth of 50 cm of water.
2. Cargo-loading computers must have a direct link to the shoreside administration.
3. The vessel must be fitted with automatic draught gauges.
4. All access points to inner compartments must be monitored by Close Circuit Television (CCTV) and have light open/shut indicators displayed to the navigation bridge.
5. Increased drainage facilities must be fitted to vehicle decks.
6. Individual units must be weighed and respective kg measured ashore for transmission to the Vessel's Cargo Officer (Figure 7.20).

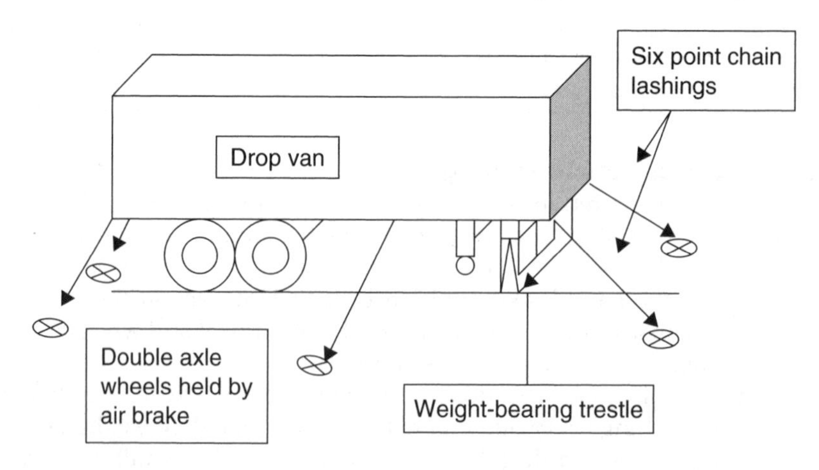

Fig. 7.20 Drop unit stowage.

Fig. 7.21 Wheeled trestle is positioned under the unit before being detached from the motor tug. Trestles fitted with spring-loaded wheels to permit easy manoeuvring under the cargo unit.

Inherent dangers associated with Ro-Ro vessels

The ships themselves generally have high freeboards and expect to experience high windage over and above the waterline. Cargo units are by necessity loaded with a high kg value, which can be detrimental to the overall metacentric height (GM). In the event of bad weather conditions, these features tend to lend to the vessel rolling heavily, which may generate units shifting.

To improve these conditions most Ro-Ro vessels are equipped with stabilizer units of either the fin varieties (fixed or deployable) or tank sluice systems or a combination of both tanks and fins. Tank systems are extremely useful when loading/discharging, as they tend to keep the vessel upright throughout cargo operations. Provided over reliance on mechanical systems does not allow complacency to permit the vessel to list over, because it is coupled with imprudent loading schedules.

If the vessel is allowed to list the vehicle ramp(s) are likely to become twisted. This may cause damage to the ramps themselves but will inevitably stop all cargo units passing over the ramps.

Passenger and cargo terminal

A Ro-Pax Ship is a Passenger Ship with Ro-Ro cargo spaces or special category spaces as defined by SOLAS Regulation II-2/3 (Figure 7.22).

Fig. 7.22 Ro-Pax ferries seen in operation at the Dover Sea Terminal.

Ship-to-shore access Ro-Ro terminal features

Link spans

Access to Ro-Ro vessels must be capable of landing vehicles at all states of tide and, in order to operate successfully, the shipboard end of the link must be able to adjust for the rise and fall of the tidal conditions prevailing. A hoist structure with associated lifting machinery is built at the shipboard end of the link to allow movement of the span to suit the rise of tide and the freeboard of respective vessels (Figures 7.23 and 7.24).

High-speed craft

The image of the Ferry World has changed considerably over the last decade. The sleek lines of mono- and multi-hull craft now operate as Ro-Pax vessels all over the world. They provide a fast and regular service mostly on the short sea trades together with some more long-haul ventures (Figures 7.25 and 7.26).

Fig. 7.23 An example of the ship-to-shore, shore-to-ship access 'link spans' which operate at the Dover Terminal. The upper enclosed passage is for passenger transit, while the lower open top links accommodate car and truck vehicle traffic.

PCCs and PCTCs

These vessels are designated to the carriage of cars. It was estimated that over 8.7 million new cars were transported in 2003, compared with 8.3 million in 2002, the main trade countries for such cargoes being Japan and South Korea. The ships are employed with multi-decks, side-loading facilities and internal ramps to facilitate high-speed-loading/discharging rates.

The ships are designed with exceptionally high freeboards and as such are susceptible to wind pressure causing considerable leeway, slowing service speed and detrimentally affecting fuel burn. More recent designs have taken this into account, and the new generation car carriers have been fitted with an aerodynamically rounded bow and bevelled along the bow-line with a view to reducing wind pressure from head winds. Six (6) PCCs operating with MOL shipping are now in service with this design feature (Figure 7.27).

Large car carriers are shipping up to 6500 car units at any one time, usually on a one-way trip, with limited prospects for return cargoes. With this in mind a high ballast capacity is generally a main feature of their operation. Where return cargoes are booked the Pure Car Carriers (PCCs) and Pure Car Truck Carriers (PCTCs) have greater flexibility.

Fig. 7.24 Link span operation. A typical link span machinery housing for hoisting and lowering the link span down to a position above the waterline. The stern ramp of the Ro-Ro ferry then lowers her stern ramp onto the links driveway to permit discharge of vehicles. The rise and fall of tide of 10.5 m is countered by the adjustment of height to the link span to allow continuous operations no matter what state of tide.

Fig. 7.25 The high-speed Ro-Pax catamaran vessel 'Millenium Dos' seen loading vehicles via the stern access, lying port side to the terminal in Barcelona, Spain.

Fig. 7.26 The Seacat Isle of Man, engaged on the Irish Sea trade between Liverpool and the Isle of Man carrying vehicles and passengers.

Fig. 7.27 The Huel Trotter car carrier manoeuvres with tug assistance fore and aft in the Port of Barcelona, Spain.

Features of the car carrier

Car carrier construction (Figure 7.28)

Typical build features:

Gross tonnage	60 587 GT	Panamax-sized vessel. Serviced by ship-to-shore
Draught	9.82 m	ramps, one at the stern (Starboard Quarter)
Air draught	52.0 m	the other midships (beam on).
Length O/A	121.08 m	Also has an option to carry refrigerated
Breadth	32.23 m	cargo on decks 5, 6 and 7 instead of doing
Service speed	21 kt	the return voyage in ballast.

The multi-deck configuration of the car carrier is in itself a striking constructional feature, the decks being interlinked by a fixed internal ramp system and elevator to lower holds. Rates of movement of car units vary directly with design but 1000 car equivalent units (CEUs) per eight (8) hour shift would not be unusual, the vessel turning round from empty, in a 48-h period.

Some decks are set at different heights to allow different head vehicles to be carried, particularly relevant where high-sided trucks may become an optional cargo. Other features of the same deck might also include higher and heavier structure to cater for the heavy-weight wheeled load. Some designs incorporate hoistable car decks offering alternative head room, as an added feature, providing additional flexibility to maximize cargo load.

A vehicle cargo mix tends to offer more options to shippers as well as being convenient in permitting direct dealing with a single carrier, the speed of cargo operations being a direct influence for shippers and on the

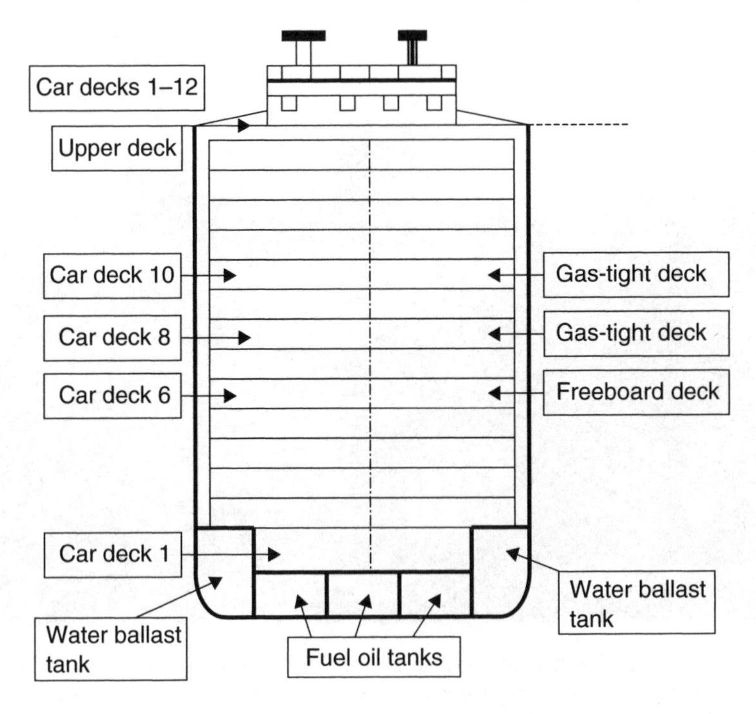

Fig. 7.28 Car carrier construction.

ship's running costs. Loading and discharge are generally achieved by a minimum of two vehicle ramps, one about the midships area while a side loading, quarter ramp, has become a popular feature of many car carriers and/or PCCs and PCTCs.

Fixed deck loading is usually about 2 tonne/m^2 throughout, though this may vary where hoistable decks are engaged. Decks are fitted with forced ventilation fan systems to clear exhaust fumes during loading and discharge periods.

As stated previously, the main disadvantage of these ships is in their construction, producing very high-sided vessels which are subject to massive wind effect when in open aspect sea conditions. As such, they experience considerable leeway which can generate increased fuel burn over a passage. Some efforts in design features, like the rounding of the bow area and bevelled bowlines, has been incorporated in some of the latest builds in an effort to increase fuel efficiency.

The ships tend to be fitted with a high ballast capacity because of the designated trade not lending to full return cargoes per voyage, although some mutual exchange cargoes that are suitable for the design decks – like palletized cargo/fork lift or tractor loading – can sometimes be arranged (Figures 7.29 and 7.30).

The new car trade is generally predominant from South Korea, Japan, Europe and Scandinavia, routing to the Canadian, USA, European and

Fig. 7.29 A specialist unit load system. The 'Republic Di Genova' one of Grimaldi's car carrier vessels seen in the Falmouth Dry Dock.

Fig. 7.30 A specialist unit load system. The angled quarter ramp and access point to vehicle decks of the above car carrier. These angled ramps have become a popular feature of car carrier vessels and are often employed with a midsection shell door ramp to improve the speed of load and discharge operations.

Australian markets. The main car carrier companies are MOL, Hual, Grimaldi and Wallenius Wilhelmsen being amongst the largest companies operating PCCs and PCTCs.

> ***Note:*** *Car carriers do not conform to conventional Ro-Ro regulations.*

The BACAT: BArge CATamaran

A double-hull catamaran shaped vessel which accommodates barges of up to about 140 tonnes. Barges are floated in from the stern and lifted from the water tunnel between the hulls by an elevator system. Additionally the 'Lighter Aboard SHip' (LASH) barges (375 tonnes) can be transported by the water tunnel with a stern door being closed up after the completion of loading.

The LASH system

A lift-on, lift-off system where lighters are raised to the upper deck by means of a moveable 'gantry crane'. They are often loaded into holds or on deck in a similar manner to containers. Alternatively, they are operated on a similar principle as the floating dock, where the parent vessel is ballasted down and the lighters are floated in via the stern, between the high-sided bulkheads. As the vessel de-ballasts, the barges are lifted into the transport.

The SeaBee: Sea barge

This system uses barge units of about 800 tonnes deadweight which are floated towards a stern elevator. An automatic transporter rolls under the barge, when at the required deck level, it is carried forward to the desired stowage position.

> ***Note:*** *LASH and SeaBee systems an also accommodate the carriage of containers.*

Chapter 8
Containers and containerization

Introduction

The first recognized container vessel was a converted World War II Tanker, named the 'Ideal X' and owned by Pan Atlantic. Her first container voyage shipped 58 containers on specially rigged decks from Port Newark, New Jersey in April 1956. Malcom P. Maclean (1914–2001) a liner-shipping pioneer, was probably the accepted founder of containerized traffic. He received the 'Admiral of the Ocean, Sea Award' in 1984 from President Reagan and 'Lloyds List' nominated him as one of the three most influential men of the twentieth century, alongside Aristotle Onassis and Ted Arison.

The first fully 'Cellular Container Ship' was a converted cargo vessel, the 'Gateway City', altered to carry 225 container units of 35 ft size. Her maiden voyage was between the Mexican Gulf and Puerto Rico but dock labour refused to work the vessel and the ship returned to the USA with her cargo.

Then the first transatlantic container line was started in 1966, and as they say, the rest is history. Door-to-door service met a huge customer demand and revolutionized the shipping industry. Containerization has all but obliterated general cargo handling, as the industry once new it. By the twenty-first century, nearly every commodity, apart from bulk products and heavy lifts, could be 'stuffed' into a container.

The largest container ships are currently being built to carry just under 10 000 TEU, and it must be anticipated that this barrier will soon be broken and even larger vessels will join the world's fleets. The system brought with it sister operations, like the Roll-on, Roll-off (Ro-Ro) system (see Chapter 7) which dovetailed with transhipping operations to feed the major terminals. Both sectors of the industry thrive today as main line contributors to cargo movement.

List of relevant container definitions and terms

Administration – means that Government of a Contracting Party, under whose authority containers are approved.

Approved – means approved by the administration.

Approval – means the decision by the administration that a design type or a container is safe within the terms of the present convention.

Cargo – is defined by any goods, wares, merchandize and articles of every kind whatsoever carried in the containers.

Cell – defined by that space which could be occupied by a single vertical stack of containers aboard a container vessel. Each stowage/hatch space would contain multiple cells, each serviced during loading/discharging by 'cell guides' (Figure 8.1).

Fig. 8.1 Empty cell guides numbered odd to starboard and even to port, situated at the fore end of the cargo container hold.

Cell guide – a vertical guidance track which permits loading and discharge of containers in and out of the ships holds, in a stable manner.

Container – is defined as an article of transport equipment: (a) of a permanent character and accordingly strong enough to be suitable for repeated use; (b) specially designed to facilitate the transport of goods, by one or more modes of transport, without intermediate reloading; (c) designed to be secured and/or readily handled, having corner fittings for these purposes; (d) of a size such that the area enclosed by the four outer bottom corners is either: (i) at least 14 m^2 (150 ft^2) or (ii) at least 7 m^2 (75 ft^2) if it is fitted with top corner fittings.

The term 'container' includes neither vehicles or packaging. However, containers when carried on chassis are included.

Container spreader beam – the engaging and lifting device used by gantry cranes to lock on, lift and load containers.

Corner fitting – is defined by an arrangement of apertures and faces at the top and/or bottom of a container for the purposes of handling, stacking and/or securing.

Existing container – is defined as a container, which is not a new container.

Flexible boxship – a term which describes a container vessel designed with flexible length deck cell guides, capable of handling different lengths of containers, e.g. 20, 30 and 40 ft units.

Gantry crane – a large heavy-lifting structure found at container terminals employed to load/discharge containers to and from container vessels. Some container vessels carry their own travelling gantry crane system on board (Figure 8.2).

Fig. 8.2 Gantry cranes engage in container cargo operations over the 'Zim California' berthed in Barcelona, Spain.

Hatchless holds – are defined as a container ship design with cell guides to the full height of the stowage without separate or intermediate hatch tops interrupting the stowage.

International transport – means transport between points of departure and destination situated in territory of two countries to at least one of which the present (CSC) Convention applies. The present convention will also apply when part of a transport operation between two countries takes place in the territory to which the present convention applies.

Karrilift – trade name for a mobile ground-handling container transporter. There are many variations of these container transporters found in and around terminals worldwide. Generally referred to as 'Elephant Trucks' or 'Straddle Trucks'.

Lashing frame/lashing platform – a mobile, or partly mobile, personnel carrier which lashing personnel can work on twist-locks at the top of the container stack without having to climb on the container tops.

Maximum operating gross weight – is defined by the maximum allowable combined weight of the container and its cargo.

Maximum permissible payload (P) – means the difference between the maximum operating gross weight or rating and the tare weight.

New container – is defined as a container the construction of which was commenced on or after the date of entry into force of the present convention.

Owner – means the owner as provided for under the national law of the contracting party or the lessee or bailee, if an agreement between the parties provides for the exercise of the owner's responsibility for maintenance and examination of the container by such lessee or bailee.

Prototype – means a container representative of those manufactured or to be manufactured in a design type series.

Rating (R) – see maximum operating gross weight.

Safety approval plate – is described as an information plate which is permanently affixed to an approved container. The plate provides general operating information inclusive of country of approval and date of manufacture, identification number, its maximum gross weight, its allowable stacking weight and racking test load value. The plate also carries 'end wall strength', the 'side wall strength' and the maintenance examination date.

Stack – a term when referring to containers, which represents the deck stowage of containers in 'tiers' and in 'bays' (Figure 8.3).

Fig. 8.3 The container stack on the deck of the 'ZIM ΣANIKAH' being discharged by terminal 'gantry cranes' in the Port of Barcelona, Spain.

Tare weight – means the weight of the empty container including permanently affixed ancillary equipment.

Terminal representative – is defined as that person appointed by the terminal or other facility where the ship is loading or unloading, who has responsibility for operations conducted by the terminal or facility with regard to that particular ship.

TEU – twenty feet equivalent unit. Used to express the cargo capacity of a container vessel.

Type of container – means the design type approved by the administration.

Type-series container – means any container manufactured in accordance with the approved design type (Figure 8.4).

Fig. 8.4 Working containers. The Mediterranean Shipping Company (MSC) Sintra lies starboard side to working containers by shoreside gantries in St John's, Newfoundland container terminal. The ships own two container cranes are turned outboard to permit access by the gantries.

Loading containers

The order of loading, when the large container vessels are carrying currently up to 10 000 TEU, must be well planned and considered as a detailed operation. Planners are usually employed ashore to provide a practical order of loading, particularly important when the vessel is scheduled to discharge at two, three or more terminal ports.

Once loading in the cell guides is complete, the pontoon steel hatch covers, common to container vessels, are replaced and secured. Containers are then stowed on deck in 'stacks' often as high as six tiers. The overall height of the deck stowage container stack may well be determined by the construction of the vessel. It must allow sufficient vision for bridge watch-keepers, to be able to carry out their essential lookout duties. The stability criteria of the vessel, when carrying containers on deck, must also be compatible with the stowage tonnage below decks.

Any deck stowage requires effective securing and this is achieved usually by a rigging gang based at the terminal. As the 'stack' is built up, each container is secured by means of specialized fittings, between containers themselves and to the ship's structure.

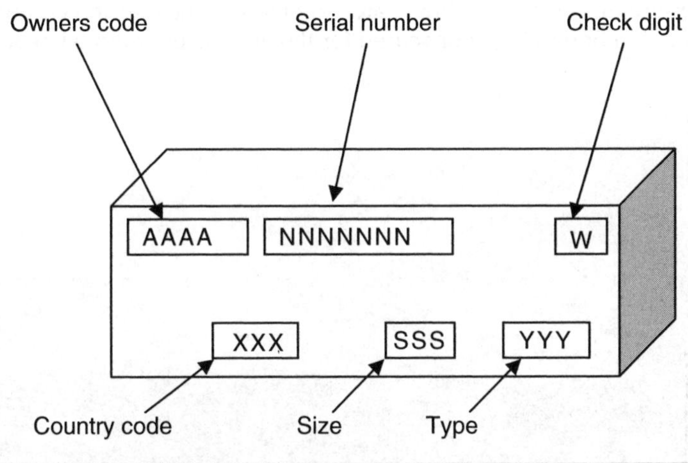

Length (m)	Width (m)	Height (m)	Gross Weight (kg)	Tare Weight (kg)	Pay Load (kg)	Usable capacity (m^3)	Imperial size (ft)
6.05	2.43	2.43	20 320.9	1590.30	18 730.6	30.75	(20′)
9.12	2.43	2.43	24 401.2	2092.92	23 308.3	46.84	(30′)
12.19	2.43	2.43	30 481.4	2593.64	27 887.0	62.92	(40′)

Fig. 8.5 Markings on containers.

Container transport

A fully laden container vessel is unlikely to be loaded down to her loadline marks despite having a container stack on deck of three or four high. Containers may weigh up to about 30-tonne gross weight each, when fully packed, but may also be empty. Hence a full capacity load may not necessarily equal the maximum permissible deadweight. If containers are carried on deck, they must be well secured by means of the iron rod lashings with associated rigging screws, fixed as part of the ship's structure. Empty or light containers could be affected by buoyancy when seas are shipped, and Deck Officers should be especially diligent when checking the upper deck stow and respective securings (Figures 8.6–8.11 and 8.19–8.22).

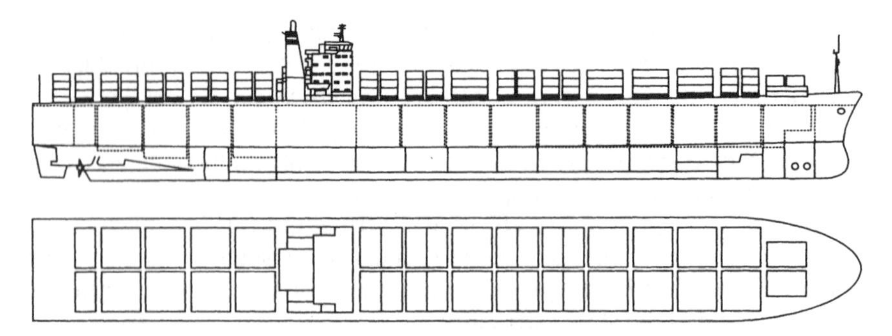

Fig. 8.6 Container vessel – upper deck stow showing stack of three high on the top of pontoon hatches. Below decks, containers secured in cell guides. The navigation bridge is not obscured for the vessels operational needs.

Fig. 8.7 The 'Dole America' container vessel manoeuvres with tugs in attendance inside harbour waters. Containers are stacked two high on deck and the two ships container cranes are seen in the stowed, fore and aft position.

Container Ship Cargo Plan

The modern type of container vessel will normally operate a container 'box' tracking system which allows continuous monitoring of any single container at any time during its transit. The plan shown in Figure 8.12 allows a six-figure number to track and identify its stowage position aboard the vessel. Distinct advantages of such a system tend to satisfy

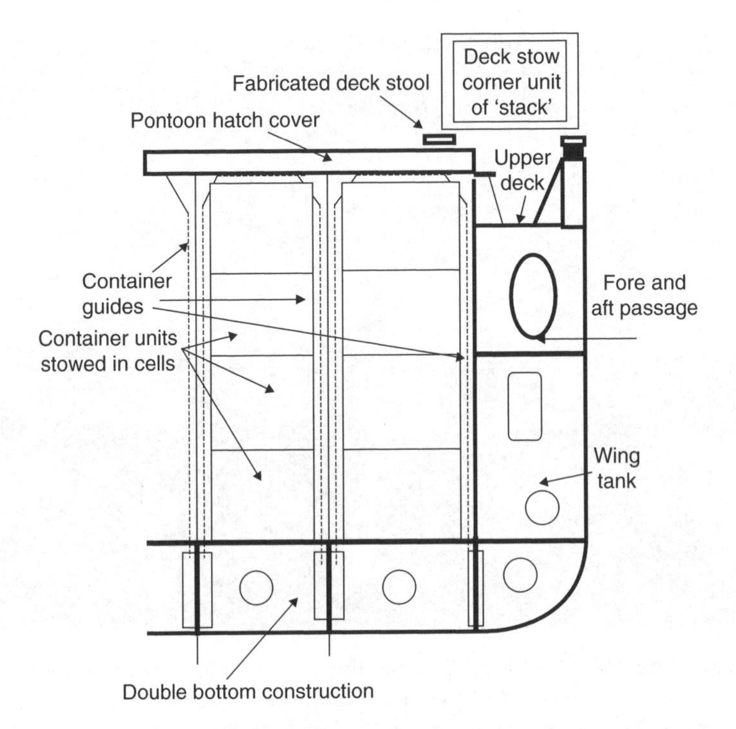

Fig. 8.8 Container vessel construction.

Fig. 8.9 The container spreader beam operates secured to the gantry crane travelling the length of the gantry jib to lift the containers on and off the vessel. The corners of the spreader beam are fitted with hinged droppable guides to ensure the beam locks can accurately locate the container corner recesses. The beam is also used to lift off pontoon hatch covers but when doing so does not deploy the hinged guides.

Fig. 8.10 The 'P&O Nedlloyd Susana' lies port side to the container terminal in Lisbon, Portugal, part loaded prior to sailing.

Fig. 8.11 The shoreside gantry cranes silhouetted against the Lisbon skyline after the P&O Nedlloyd Susana departs the berth.

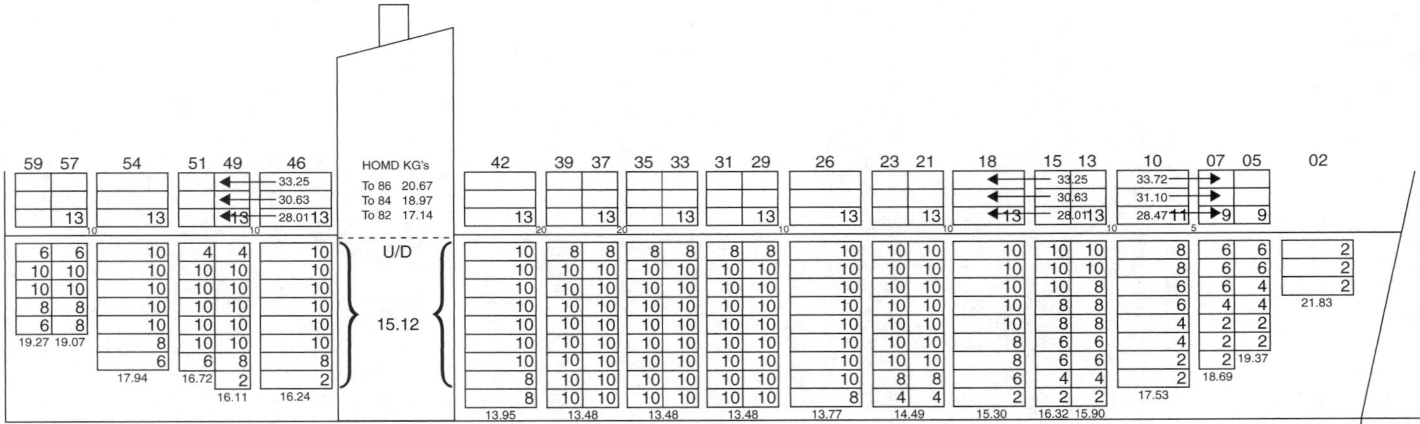

Bays	59 57	54	51 49	46	HOMD KG's: To 86 20.67; To 84 18.97; To 82 17.14	42	39 37	35 33	31 29	26	23 21	18	15 13	10	07 05	02	Total
1 High on deck	26	13	26	13	1 High on deck	13	26	26	26	13	26	13	26	11	18	Nil	200 × 20 76 × 40
Under deck P	20 21	27	30 32	35	P Under deck	43	44 44	44 44	44 44	44	41 41	37	32 31	20	14 12	3	1076 × 20
Under deck S	20 21	27	30 32	35	S Under deck	43	44 44	44 44	44 44	44	41 41	37	32 31	20	14 12	3	418 × 40
Total	108	67	150	83	Total	99	202	202	202	101	190	87	152	51	70	6	1276 × 20 494 × 40

The modern type of container vessel will normally operate a container 'box' tracking system which allows continuous monitoring of any single container at any time during its transit. The plan allows a six figure number to track and identify its stowage position aboard the vessel. Distinct advantages of such a system tend to satisfy shipper enquiries as well showing that the shipping company is efficient in its business.

Other aspects of secutiry are also clearly beneficial in a security conscious age.

An example tracking system could be typically:

The first two numbers of the six-digit number	= The identification of the 'Bay' of stowage.
The second two numbers	= The 'cell' of stowage
The last two numbers	= The level/tier of stowage

Fig. 8.12 Container Ship Cargo Plan.

shipper enquiries as well as showing that the ship company is efficient in its business. Other aspects of security are also clearly beneficial in a security conscious age. An example tracking system could be typically: the first two numbers of the six-digit number (the identification of the 'bay' of stowage); the second two numbers (the 'cell' of stowage) the last two numbers (the 'level tier' of stowage).

Container types

There are many container types in operation to suit a variety of trades and merchandize. Sizes also vary and they can be shipped in the following sizes: 8 ft in width and 8 ft or 8 ft 6 inch in height, with lengths of 10, 20, 40 or 45 ft.

Conventional units (general purpose) – also known as a dry container are made from steel and fully enclosed with a timber floor. Cargo-securing lashing points are located at floor level at the base of the side panelling. Access for 'stuffing' and 'de-stuffing' is through full height twin locking doors at one end.

Open top containers – covered by tarpaulin and permits top loading/discharging for awkward sized loads which cannot be easily handled through the doorways of general purpose containers. These may be fitted with a removable top rail over and above the door aperture.

Half-height containers – an open top container which is 4 ft 3 inch in height, i.e. half the standard height of a general purpose container. They were designed for the carriage of dense cargoes such as steel ingots, or heavy-steel cargoes or stone, etc. since these cargoes take up comparatively little space in relation to their weight, two half-height containers occupying the same space as the standard unit.

Flat rack container – this is a flat bed with fixed or collapsible ends and no roof. They are used to accommodate cargoes of non-compatible dimensions or special cargoes that require additional ventilation.

Bulk container – are containers designed to carry free flowing cargoes like grain, sugar or cement. Loading and discharging taking place via three circular access hatches situated in the roof of the unit. They also incorporate a small hatch at the base which allows free flow when tipping the unit. Such containers are usually fitted with steel floors to facilitate cleaning.

Tank containers – are framed tank units designed for the carriage of liquids. The cylindrical tank usually made of stainless steel is secured in the framework which is of standard dimensions to be accommodated in loading and discharging as a normal general purpose container unit. The tanks can carry hazardous and non-hazardous cargo and are often used for whisky or liquid chemicals.

Ventilated containers – generally designed as a general purpose container but with added full length ventilation grills at the top and bottom of the side walls of the unit. They were primarily designed for the coffee trade but are equally suitable for other cargoes, which require a high degree of ventilation during shipping.

Open-sided containers – these units are constructed with removable steel grate sides which are covered by poly vinyl chloride (PVC) sheeting. The side grates allow adequate ventilation when it is used to carry perishable goods and/or livestock. Such containers permit unrestricted loading and discharging with the grates removed.

Insulated containers – are insulated and often used in association with a refrigeration air-blower systems to keep perishable cargoes fresh, e.g. meats, fruits vegetables, etc. The container has two porthole extractors fitted to one end of the unit to allow the cool air circulation to operate from the cooling plant. They are generally stowed under deck and close to, or adjacent to, the ship's circulation ports. Other types of containers in this category rely only on the insulation and are not fitted with cooling plant, and these can be stowed in any position on the ship.

Refrigerated containers – more generally known as the reefer container, they are totally insulated and fitted with their own refrigeration plant. They must be connected to the ship's mains and require close stowage to a situated power point. They are usually employed for holding foodstuffs, meat and dairy products being prime examples. These units have become prolific and have caused a major reduction in the numbers of dedicated 'reefer ships', although reefer ships still operate they tend to be limited to specific trades like 'bananas' (Figures 8.13 and 8.14).

Fig. 8.13 The 'OOCL Shanghai' lies port side to the container terminal in Barcelona, Spain, after completing cargo loading with a full container load, the deck stack being at a six-tier height. The terminal 'gantry cranes' seen in the upright and clear position (the ship is not fitted with its own cranage).

Fig. 8.14 Container (internal) hatch stowage. Container hatch with part load containers lying in the cell guides of the lower hold cargo space.

Reefer containers

With many of the chilled and frozen products being transported by sea containers there was bound to be an influence on the reefer trade; so much so that designated 'reefer' ships have been greatly reduced in number, other than possibly in the banana trade. Ro-Ro units, as well as the specified refrigerated containers, have now dominated the reefer commodity shipping markets.

The container units themselves are built with insulation and pre-cooled prior to being loading at the handling station, a shore power supply being used to activate the units cooling plant. Once packed and sealed the temperature of the unit is lowered to the desired level and monitored by a temperature sensor attached to the container. As soon as the unit is packed, the refrigeration machinery is activated either by the continued use of a shore supply or linked directly to the transporters (mobile) power source.

Terminals and container parks have specialized park areas to enable mobile units to switch to a static shore power supply, once the mobile transport supply is stopped. Disconnection of units takes place just prior to loading on board the ship. The supply is reconnected from the ship's mains once the unit is stowed in its allocated position aboard the vessel.

The modern container vessel can expect to carry numerous units with refrigerated cargoes, all plugged into the ship's power supply fitted to specified loading bays. They would, in the main, be fitted with a reefer container monitoring system to ensure that temperatures are retained within acceptable limits.

'Reefer' container monitoring

Various types of monitoring systems are available for shipping operators, either stand-alone or integrated operations which could include tank gauge systems, ballast control, power management, fire fighting, etc.

The local control unit indicated could monitor up to 3000 cargo units, or numerous tanks for pressure, temperature, volume, viscosity, etc. (Figure 8.15).

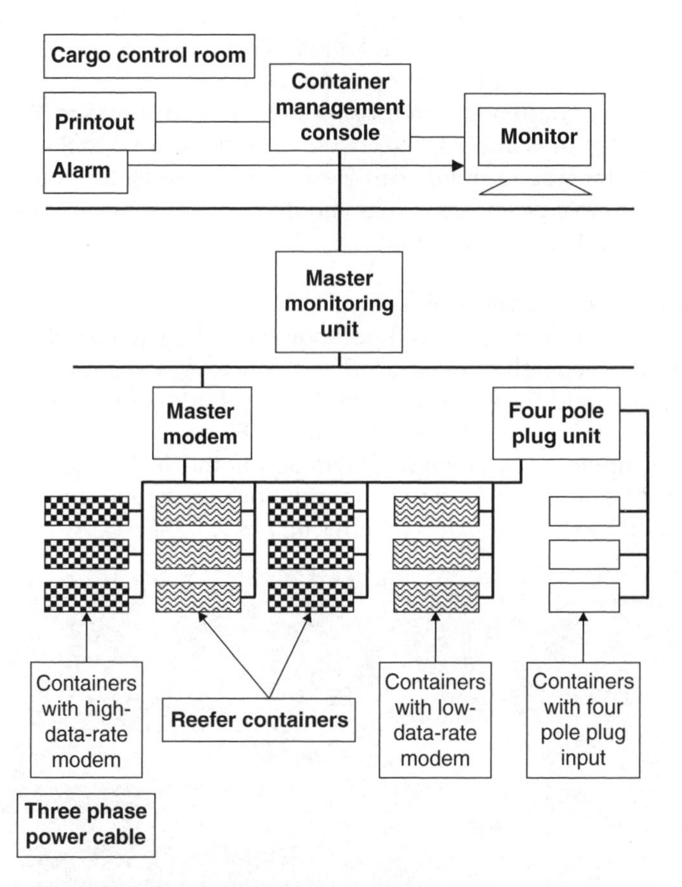

Fig. 8.15 Cargo control room – container monitoring.

Containers on deck

It is regular practice to carry containers on deck on both designated container vessels and general cargo/service vessels. Further recommendations on deck stowage are advised by 'M' Notice 1167.

Deck containers should be stowed and secured taking account of the following:

1. Containers should preferably be stowed in a fore and aft direction.
2. They should be stowed in such a position as not to deny safe access to those personnel necessary to the working of the vessel.

3. They should be effectively secured in such a manner that the bottom corners will be prevented from sliding and the top corners will be restrained to prevent tipping.
4. The unit should stowed in a manner that it does not extend over the ships side (many containers are stowed part on the hatch top and part on extending pedestal supports, but the perimeter of the unit is kept within the fine lines of the vessel).
5. Deck containers should be carried at a single height (one high). However, this may be increased if twist-locks are used to secure the bottom of the container to a fabricated deck stool.
6. Deck loads should not overstress the deck areas of stowage. Where units are on hatch tops, these hatch covers must be secured to the vessel.
7. No restraint system should cause excessive stress on the container.
8. Restraint systems and securings should have some means of tightening throughout the voyage period.

Container deck stowage

Container decks, and reinforced pontoon hatch tops to take the deck load capacity, are generally constructed with increased scantlings to satisfy Classification and Construction Regulations. Both open decks as seen in Figure 8.16 and the pontoon hatch cover (Figure 8.23), when fitted, are usually equipped with container feet to permit the 'boxes' to be locked into position. The first tier, being the foundation for second and subsequent tiers, would be stowed on top (Figures 8.17–8.22).

Fig. 8.16 The exposed container cargo deck of the 'Baltic Eider' seen with the container deck stool-securing points in uniform rows to form the basis of an even stow.

Fig. 8.17 Part loaded deck of the 'Sete Cidades'. Containers seen on the hatch tops as the vessel lies starboard side to the container terminal in Oporto. The ship's own two container cranes are turned outboard to allow access to the shoreside gantry cranes.

Fig. 8.18 Part loaded deck of the vessel 'Hydra J'. Containerized vehicles and half-height container seen clear of securings and ready for discharge.

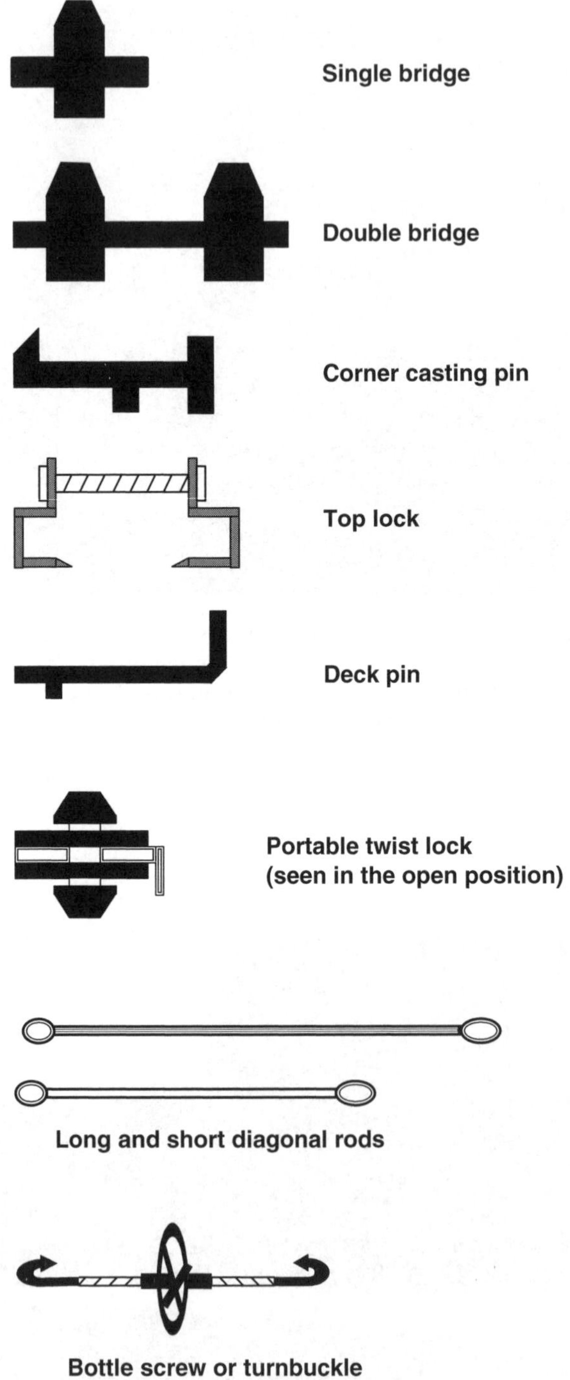

Fig. 8.19 Container lashing fitments.

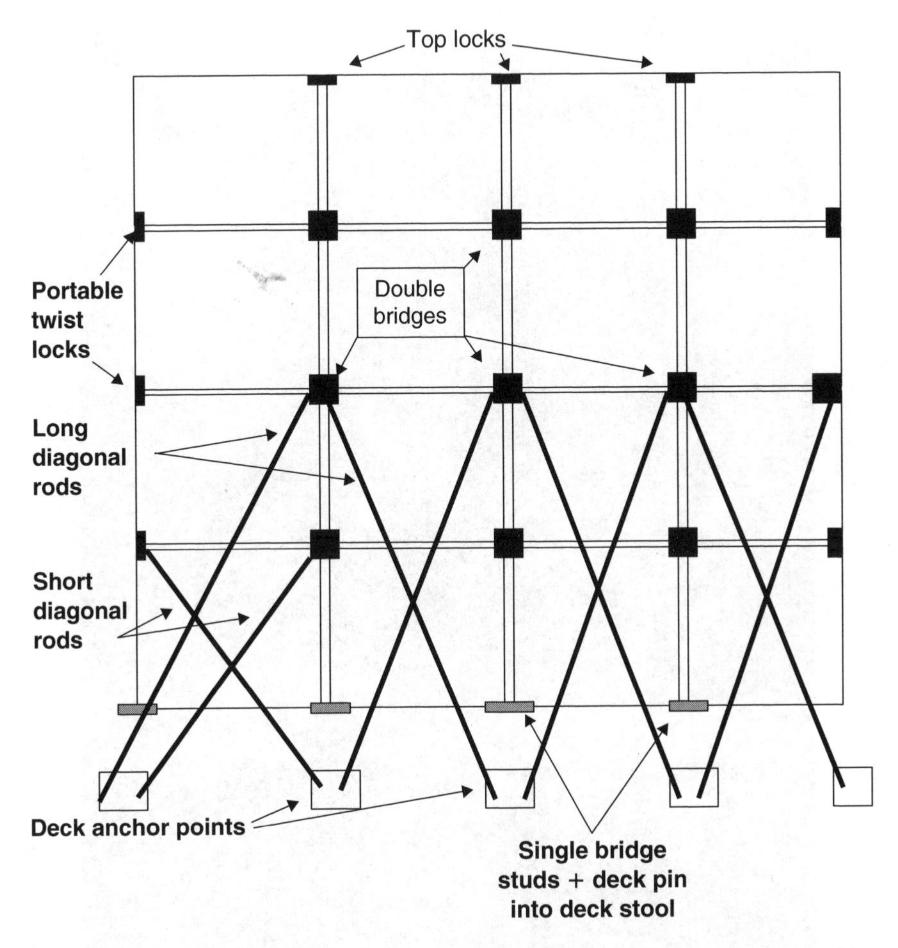

Fig. 8.20 Container deck stowage example. Short and long rods secured by bottle screw or turnbuckle to deck anchor points.

Loadicator and loading plan computers

Many ships are now equipped with loadicator systems or a loading computer with appropriate software. It is usually a conveniently sited visual display for the Master and the Loading Officers and is gainfully employed on Ro-Ro vessels, container ships, tankers and bulk carriers. The system should ideally be interlinked with the shoreside base to enable data transmissions on, unit weights/tonnages/or special stow arrangements.

The computer would permit the location and respective weights of cargo/units to be entered quickly and provide values of limiting measured distance between the keel and the centre of gravity of the vessel (KG) and 'metacentric height' (GM) together with deadweights at respective draughts/displacements. It would also have the capability to provide a

Fig. 8.21 Deck stowage and securing of container stack. The deck stow is seen in way of the edge of the pontoon hatch cover. The lashings are to the base, third tier container level by long bar (not shown). The second tier container level being secured by short bar lashings with the base containers locked to the pontoon.

printed record of the state of loading and show a visual warning in the event of an undesirable stability condition or overload occurring.

Distribution of the ship's tank weights, stores and consumables affecting final calculations, and total displacement would also be identifiable within the completed calculations. The primary aim of the loading computer is to

Fig. 8.22 Container lashing bars, seen secured from the main deck to the first and second tier of deck containers.

Fig. 8.23 The container hoist engaged in lifting the pontoon hatch covers clear of the cellular holds.

Fig. 8.24 Tracked shoreside 'gantry crane' for the loading of containers, semi-automatic, driverless transports deliver containers to its underside for loading aboard the container vessel.

Fig. 8.25 Shoreside container terminal showing automatic stacking cranes in the background.

ensure that the vessel always departs the berth with adequate stability for the voyage. If this situation can be achieved quickly, costly delays can be eliminated and safety criteria is complied with.

The data required to complete the stability calculations would need to be supplied by the shoreside base with regard to cargo weights. This in turn would be certificated by the driver – for Ro-Ro unit loads – obtaining a load weight certificate authorized from an approved 'weight bridge' prior to boarding the vessel. Draught information would inevitably come from a 'Draught Gauge System' for the larger vessel and be digitally processed during the period of loading.

A ship's personnel could expect to become familiar with manipulation of the changing variables very quickly alongside the fixed weight distribution throughout the ship. This would permit, in general, few major changes to the programme, especially on short sea ferry trade routes where limited amounts of bunkers, water and stores are consumed and values stay reasonably static.

Fixed weights are applicable to a variety of units or vehicles and, as such, where units are pre-booked for the sea passage, an early estimate of the ship's cargo load, and subsequent stability, can often be achieved even prior to the vessels arrival.

The loadicator programmes provide output in the form of:

- shear forces and bending moments affecting the vessel at its state of loading
- cargo, ballast and fuel tonnage distributions
- a statement of loaded 'GM', sailing draughts and deadweight.

Terminal operations

The sheer size of 'container terminals' around the world must generate cause for the tremendous volume of work which is employed in the transport, storage and shipping of the many units. The general public would only visibly see the number of units which a terminal has inside its perimeter at any one time. However, the maintenance of the gantry cranes, the ground-handling transports, the documentation concerning a single 'box unit' become the invisible operations that generate a successful terminal. They employ considerable manual workers with various skills, from the wharf men to personnel engaged in 'stuffing container units' – security personnel, administration staff, maintenance workers, ships planners, etc. not to mention the insurance and legal professionals engaged in the background.

The largest terminals in the world are shown in Table 8.1.

The increased growth of unit movements is based on figures from 2002/2003 and should not be considered for future years, as world trade is influenced by many factors; not least the strength of national economies,

Table 8.1 Main terminals throughout the world

Terminal port	*Number of TEU handled (millions)*	*Annual percentage increase (+/–)*
Hong Kong	20.82	+8.8
Singapore	18.41	+8.7
Shanghai	11.37	+32.1
Shenzhen	10.65	+39.9
Pusan	10.37	+9.7
Kaohsiung	8.81	+3.8
Rotterdam	7.10	+9.2
Los Angeles	6.61	+8.4
Hamburg	6.14	+14.2
Antwerp	5.44	+14.0

the strength of the US dollar, the emergence of China and Charter rates to mention but a few of the relevant influences.

However, what is clear is that if the location of the ports is noted then the geography would indicate that the USA, Europe, the Far East, and, in particular, China are emerging as the main trading blocks for containerized traffic. Feeder operators to Australasia, India, the Baltic and Mediterranean regions continue to flourish in support of the major operators (Figures 8.26–8.31).

Fig. 8.26 The container gantry cranes discharge the vessel 'Zim Marseille' which lies port side to the container terminal in Barcelona, Spain.

Fig. 8.27 A single gantry crane and a mobile dock side crane work the container cargo of the feeder container vessel 'Providence'. The ship's own container cranes are seen turned outboard to facilitate the shoreside loading systems in the port of Barcelona.

Fig. 8.28 Automated stacking cranes provide unit movements to the terminals container stack. The containers being loaded to driverless ground-handling transports.

Fig. 8.29 Automated, driverless ground-handling transports deliver container units from the terminal stack to the underside of ship/shore gantry cranes.

Fig. 8.30 Tracked, terminal stacking gantry cranes operate through the Lisbon container terminal. Rail and road transport having access to the container park.

Fig. 8.31 Walkways, road paths and track rails for gantry cranes seen alongside the berth, opposite the terminal stacking gantries. Container avenues lying parallel to the ships berth.

Container operations

Shipping and booking

In order to ship a container certain procedures and documentation processes are required and the freight office of a shipping company would require the following information:

1. Name and address of company booking the unit for export
2. Bill of Lading (B/L), with name and address of shipper if different from above
3. The quantity of cargo to be shipped: including weight, measurement, marks and numbers of packages
4. Name of port of discharge
5. Commodity details, hazardous, refrigeration required and/or precise description of goods
6. The place of delivery and acceptance
7. Place of packing the container
8. Earliest date of container availability
9. Customs assigned number
10. Customs status of cargo declaration.

The container would then be designated a booking reference number to allow a constant trace to be maintained on the unit while being exported.

The Export Container Packing Document: ECPD

A detailed packing list of the container is required and this serves as not only a list of container contents but also includes: (a) quarantine declaration (if required); (b) transport document for (i) receipt from shippers for empty container, (ii) receipt from shipping company for full container; (c) stated conditions which relate to the use of equipment at shippers premises and (d) Declaration of Customs Status.

B/L

The shipping company will produce a 'B/L' once they have been informed of all relevant details regarding the nature of the cargo. It would be normal practice that a 'freight invoice' would also be issued at this time, as the B/L and the freight invoice are both computer generated.

Shipping procedures

Figure 8.32 shows the likely procedure that would be followed in order for a smooth outcome to ensue. It is normal practice for the shipper to have a financial bond in place prior to shipment, the function of the bond being to guarantee payment at the country of entry of the goods.

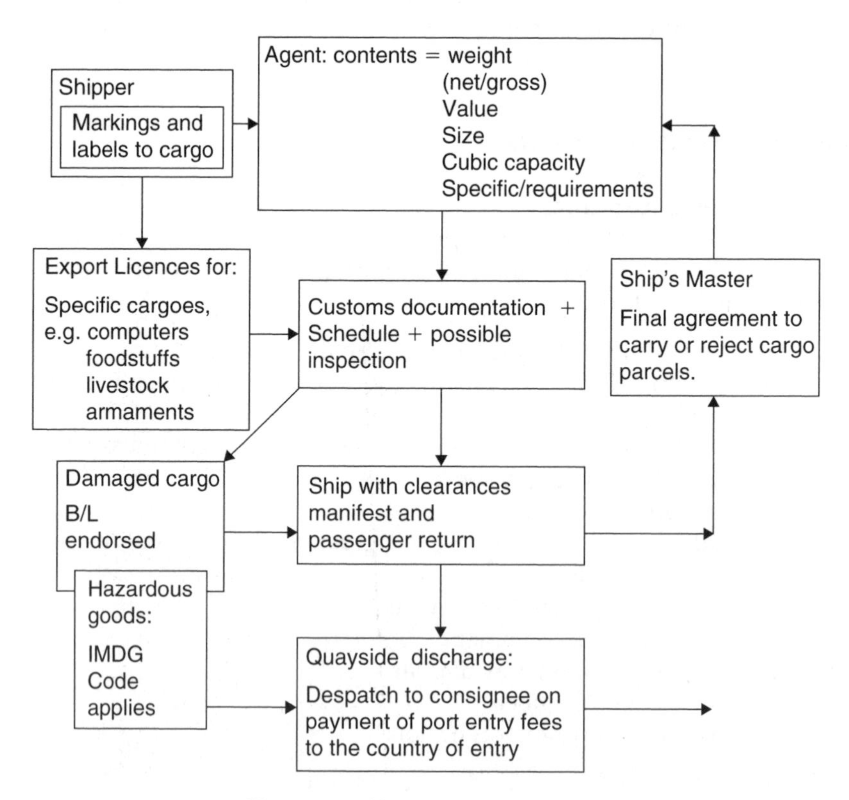

Fig. 8.32 Shipping procedures.

Fig. 8.33 Ground handling of containers – equipment and methods.

Fig. 8.34 Lifting, stacking and transporting.

Fig. 8.35 Container terminal movement. Container transporter, with lift, transport, and stacking capability. A type of container karrilift based on the fork lift truck principle which is widely employed in container terminals worldwide.

Chapter 9

Special cargoes, hazardous goods and deck cargoes

Introduction

All cargoes must be considered special in some way, particularly so, if it is the first time that an individual has had experience of that specific cargo. It would take a lifetime for a mariner to carry every commodity and even then, certain products would be absent from the list. The interpretation of special cargoes can encompass many types of cargoes but it is generally accepted that those parcels that require special or additional attention for their safe transport and discharge, fall into this category.

Clearly, hazardous goods covered by the International Maritime Dangerous Goods (IMDG) Code are deeply entrenched under this particular umbrella of special cargoes. However, it is not just hazardous materials. Valuables, like bullion, bank notes, stamps, or personnel effects requiring lock-up stowage conditions, are also considered as specials.

Within the scope of this chapter falls 'deck cargoes'. They are exposed to the elements and are often the first to suffer from any misadventure which may befall the ship's voyage. Deck cargoes, by their very nature, may fall into the class of hazardous goods or they may, like timber deck cargoes, have their own inherent dangers, which may threaten the well-being of both ship and cargo.

Whatever goods are shipped, it is essential that correct stowage procedures are taken from the onset. They should be clearly noted on the stowage plan and relevant persons should be made aware of the nature of potential hazards or special precautions that should accompany the transport.

Definitions and terminology relating to hazardous cargoes

Auto-ignition temperature – is the lowest temperature at which a substance will start to burn without the aid of an external flame. Spontaneous combustion begins, provided that conditions are right, when auto-ignition temperature is attained.

Carrier – means any person's organization, or government, undertaking the transport of dangerous goods by any means of transport. This includes both carriers for hire or reward (known as common or contract carriers) and carriers on own account (known as private carriers).

Control temperature – means the maximum temperature at which certain substances (such as organic peroxides and self-reactive and related substances) can be safely transported during a prolonged period of time.

Cylinders – are transportable pressure receptacles of a water capacity not exceeding 150 l.

Dangerous goods – means substances, materials and articles covered by the IMDG Code.

Defined deck area – means that area of the weather deck of a ship or of a vehicle deck of a Roll-on, Roll-off (Ro-Ro) ship which is allocated for the stowage of dangerous goods.

Emergency temperature – means that temperature at which emergency procedures shall be implemented.

Flammable liquid – is a liquid having a flash point lower than 37.8°C. A combustible liquid is a liquid having a flash point of 37.8°C or above, e.g. gasoline is a flammable liquid, whereas kerosene is a combustible liquid.

Flammable range – the limits of flammable (explosive) range, in the range between the minimum and the maximum concentrations of vapour in air which forms a flammable (explosive) mixture. Usually abbreviated to LFL (lower flammable limit) and UFL (upper flammable limit). These are synonymous with the lower and upper explosive limits.

Flash point – is that lowest temperature at which a liquid gives off sufficient vapour to form a flammable mixture with air near the surface of the liquid, or within the apparatus used. Flash point represents the change point from safe to risk.

Harmful substances – are those substances that are identified as marine pollutants in the IMDG Code.

International Maritime Dangerous Goods Code – a mandatory code for the carriage of dangerous goods at sea as adopted by the Maritime Safety Committee (MSC) of the International Maritime Organization (IMO). Effective from 1 January 2004 this code is applicable to all ships to which the Safety of Life at Sea (SOLAS) convention applies (Resolution MSC. 122(75)).

Medical First Aid Guide – a section of the supplement to the IMDG Code which details guidelines for the application of first aid to persons exposed and affected by hazardous goods.

Packaged form – means the form of containment specified in the IMDG Code.

Settled pressure – means the pressure of the contents of a pressure receptacle in thermal and diffusive equilibrium.

Sift proof – is packaging which is impermeable to dry contents including fine solid material produced during transport.

Tank – means a portable tank (including a tank container) a road tank vehicle, a rail tank wagon or a receptacle with a capacity of not less than 450 l to contain solids, liquids or liquefied gases.

Water reactive – means any substance which in contact with water emits flammable gas.

Working pressure – means the settled pressure of a compressed gas at a reference temperature of 15°C in a full pressure receptacle.

The IMDG Code

The IMDG Code is the recognized code of practice for the carriage of hazardous cargoes and is covered by four volumes, plus a supplement.

IMDG Code Volume 1

This contains a general introduction and covers standards on the:

1. classification of goods
2. packaging of those goods
3. documentation required when shipping
4. marking labelling and placarding required
5. standards concerning explosives in passenger vessels.

Various sections cover the above standards for Classes 1–9 hazardous goods, in a more detailed format. Annex 1 follows the introduction and provides details on modes of packaging to UN standards.

An alphabetical General Index of all the substances, inclusive of the UN number, class, packaging group, follows the Annex. This index should be employed as the first step to retrieve information affecting a particular cargo substance. Included here is also the Medical First Aid section with the *Medical First Aid Guide* (MFAG) table numbers. Definitions, abbreviations and explanatory notes complete the volume.

IMDG Code Volume 2

This volume contains detailed instructions regarding the packing, labelling and stowing of explosives (including the specific requirements for the construction of magazines), together with individual schedules for substances in 'Class 1' explosives, 'Class 2' gases and 'Class 3' flammable liquids.

IMDG Code Volume 3

This volume covers 'Class 4' flammable solids and 'Class 5' oxidizing agents and organic peroxides. Each schedule contains specific instructions on the packaging and stowing, and relevant information regarding each class.

IMDG Code Volume 4

This volume covers Classes 6 (poisons), 7 (radioactive substances), 8 (corrosives) and 9 (miscellaneous). It also details specific information for each class including segregation details, toxicity level and radioactive rating scales, as well as radioactives for offshore supply vessels, package requirements for corrosives and miscellaneous substances.

Supplement of the IMDG Code

The supplement contains emergency procedures (EmS) and schedules for particular commodities, plus details of specialized equipment required for handling spills and fires. The MFAG provides information on symptoms and the body's reaction to exposure following an accident, as well as safe practice for handling of solid bulk cargoes, particularly concentrates. Methods of reporting procedures for vessels involved in incidents are also covered.

Shipping procedure for the loading and transport of hazardous goods

To transport dangerous goods by sea, they must pass through the following procedures:

1. The shipper is responsible for obtaining 'Export Licences' for the goods in question.
2. The shipper would also be responsible for marking and labelling the goods to be shipped in accord with the IMDG Code.
3. Following contact with the shipping company, agents must provide:
   - the number of packages together with their weight
   - the value of the goods
   - special requirements for carriage of the goods.
4. Customs clearance would be required as for any other cargo.
5. The Bill of Lading would be sighted and seen to be free of endorsements.
6. The goods would be entered on the ship's manifest and marked on the cargo stowage plan.
7. A ship's Officer would check the UN number, the details of the commodity, the labelling of the package and the condition of the packaging. Any special stowage arrangements would be noted and observed at this stage.
8. The Ship's Master has the right to accept or reject the cargo prior to loading.

Once the goods are stowed on board the vessel the requirements of the IMDG Code would be followed throughout the period of the voyage.

> **Note:** *Reference should also be made to Annex III of Maritime Pollution (MARPOL), regarding the Regulations for the Prevention of Pollution by Harmful Substances, carried at sea in packaged form.*

If appropriate, a 'Document of Compliance' for the carriage of certain hazardous goods may be required by the ship.

Documentation for shipping dangerous goods

1. Where dangerous goods are to be carried by sea, all documentation relating to the goods must carry the correct technical name where the goods are named. The use of a trade name alone must not be used.
2. Any shipping documents prepared by the shipper must include or be accompanied by a signed certificate or declaration that the shipment offered for carriage is correctly packaged and marked, labelled, etc. and is in proper condition for shipment.
3. The person responsible for the packing of dangerous goods in a freight container or road vehicle must provide a signed container packing certificate or a vehicle packing declaration, which states that the cargo in the unit has been correctly packed and secured and that all applicable transport requirements have been fulfilled.
4. In the event that a freight container or road vehicle containing dangerous goods is not compliant with the above, then such vehicle or container shall not be accepted for shipment.
5. Every ship carrying dangerous cargo shall have a special list or manifest of such dangerous goods on board contained within a detailed stowage plan. Such documents will identify by class and location all such dangerous goods on board the vessel. Copies of these documents will be available prior to departure to a person as designated by the Port State Authority.
6. In the case of marine pollutants, the signed shipping documents must also state that the parcel offered for shipment is a marine pollutant and that as such it is in a proper condition for carriage by sea.

> ***Note:*** *A copy of the stowage plan must be retained ashore until the harmful substances have been discharged from the vessel.*

Documentation detail – for shipping dangerous goods

One of the prime functions of any documentation that accompanies dangerous goods for shipping is to provide basic information associated with the hazardous substance. To this end, the shipping document for each product, material or article offered for shipment must include the following:

1. The proper shipping name.
2. The class and when assigned, and the division of the goods.
3. The UN number.
4. The packaging group for the substance carried under a 'Not Otherwise Specified' (NOS) notation or other generic entry which may include the possibility of the assignment of more than one packaging group.
5. For 'Class 7', radioactive materials only, the Class 7 schedule number.
6. Any empty or any packages containing residual dangerous goods must be marked by the words *empty uncleaned* or *residue-last contained,* before or after the proper shipping name of the substance.

7. Where dangerous goods waste (except radioactive waste) are being transported for disposal, the proper shipping name should be proceeded by the word *waste*.
8. The number and kind of packages together with the total quantity of dangerous goods covered by the description.
9. The minimum flash point if 61°C or below (°C closed cup test), or other additional hazard which is not communicated in the description of the dangerous goods.
10. The identification that the goods are *marine pollutants* and when declared under an NOS, or generic entry, the recognized chemical name of the marine pollutant in parentheses.
11. For Class 4.1 self-reacting substance or a Class 5.2 organic peroxide, the control and emergency temperatures, if applicable.

Additional information is required where special classes of dangerous goods are carried and this information is applicable for: all 'Class 1' goods, gases, infectious substances, radioactive materials, certain substances in Class 4.1 which may be exempt from display of an explosive subsidiary label, and certain organic substances that are also exempt from displaying the explosive subsidiary label.

Classes of dangerous goods

Dangerous goods are classified as follows:

Class 1 Explosives

Class 2 Flammable gases, poisonous gases, or compressed, liquefied, or dissolved gases which are neither flammable nor poisonous

Class 3 Flammable liquids, subdivided into three categories:

- 3.1 Low flash point group of liquids having a flash point below −18°C (0°F) closed cup test, or having a low flash point in combination with some dangerous property other than flammability
- 3.2 Intermediate flash point group of liquids having a flash point −18°C (0°F) up to but not including 23°C (73°F) closed cup test
- 3.3 High flash point group of liquids having a flash point of 23°C (73°F) up to and including 61°C (141°F) closed cup test

Class 4

- 4.1 Flammable solids
- 4.2 Flammable solids or substances liable to spontaneous combustion
- 4.3 Flammable solids or substances which in contact with water emit flammable gases

Class 5

- 5.1 Oxidizing substances
- 5.2 Organic peroxides

Class 6
6.1 Poisonous (toxic) substances
6.2 Infectious substances.
Class 7 Radioactive substances
Class 8 Corrosives
Class 9 Miscellaneous dangerous substances. That is, any other substance which experience has shown, or may show, to be of such a dangerous character, that this class should apply to it.

Stowage of Class 1: explosives

Explosives are categorized for stowage in one of the following methods:

1. Stowage Category I – goods not requiring a magazine stowage.
2. Stowage Category II, Type 'A' – a fixed magazine structure. This magazine should be close boarded on the inner sides and floor. Although cargo battens are sufficient on the ship's sides and bulkheads if they are not more than 150 mm apart.
3. Stowage Category II, Type 'B' – fixed magazine structure. Similar to 'Type A' but close boarding of sides and floor is not a requirement.
4. Stowage Category II, Type 'C' – a fixed magazine structure similar to 'Type B', but restrictions are placed on the permitted distance from the ship's side.
5. Stowage Category II – approved portable units.
6. Stowage Category II – freight containers.
7. Stowage Category III (pyrotechnics) – similar stowage to Category I, except that goods should not be overstowed with other cargo.
8. Stowage Category IV – the goods requiring this stowage should be placed as far as possible away from living accommodation and should not be overstowed. Deck stowage is preferred.

Package requirements for dangerous goods

All dangerous goods intended for carriage by sea must conform to the specifications and performance tests as recommended by the IMDG Code.

Packaging must be:

1. well made and in good condition,
2. sealed to prevent leakage,
3. of a package material which should not be adversely affected by the substance it is containing within. If necessary it should be provided by an inner coating capable of withstanding ordinary risks of handling and carriage by sea. Where the use of absorbent material or cushioning material is employed, that material shall be:
   - capable of minimizing the dangers to which the liquid may give rise,
   - so disposed as to prevent movement and ensure that the receptacle remains surrounded,
   - where reasonably possible, of sufficient quantity to absorb the liquid in the event that breakage of the receptacle occurs.

When filling packages/receptacles with liquids, sufficient ullage should be left to make an allowance for expansion which may be caused by rises in temperature.

Gas cylinders for gases under pressure must be adequately constructed and tested, maintained and correctly filled. When pressure may develop in a package by the emission of gas from the contents due to a rise in temperature, such a package may be fitted with a vent, provided that the gas emitted will not cause danger in any form to the surround.

Marking of dangerous goods (Ref. IMDG Code)

Packages of 'dangerous goods' must be transported in accordance with the provisions of the IMDG Code. Packages containing a harmful substance should be durably marked with the correct technical name (trade names alone should not be used). They should be marked to indicate that they are a marine pollutant and identified by additional means like by use of the relevant UN number.

Markings on packages containing harmful substances must be of such a durable nature as to withstand three (3) months immersion in sea water. They must be adequate to minimize the hazard to the marine environment having due regard to their specific contents.

> **Note:** *Packages that contain small quantities of harmful substances may be exempt from the marking requirements. Exemptions are referenced in the IMDG Code.*

Empty packages which have previously been used for the transport of harmful substances shall themselves be treated as harmful substances, unless adequate precautions have been taken to ensure that they contain no residues that are of a harmful nature to the marine environment.

Purpose of marking and labelling

The purpose of marking packages with the correct and proper shipping name, and the UN number of the substance, is to ensure that the material or substance can be readily identified during transportation of the goods. This identification is particularly important in determining the nature of emergency treatment which would be required in the event of a spillage or accident occurring (Figure 9.1).

Carriage in cargo transport units

The shipper is responsible for providing the transport documents; namely a signed certificate that the unit offered for carriage is properly packaged, marked and labelled or placarded, as appropriate. If dangerous goods have been packed in such a unit and the packing certificate is not available, the cargo transport unit should not be accepted for carriage.

Segregation

Dangerous goods that have to be segregated from each other must not be transported in the same cargo transport. Exceptions to this apply and are

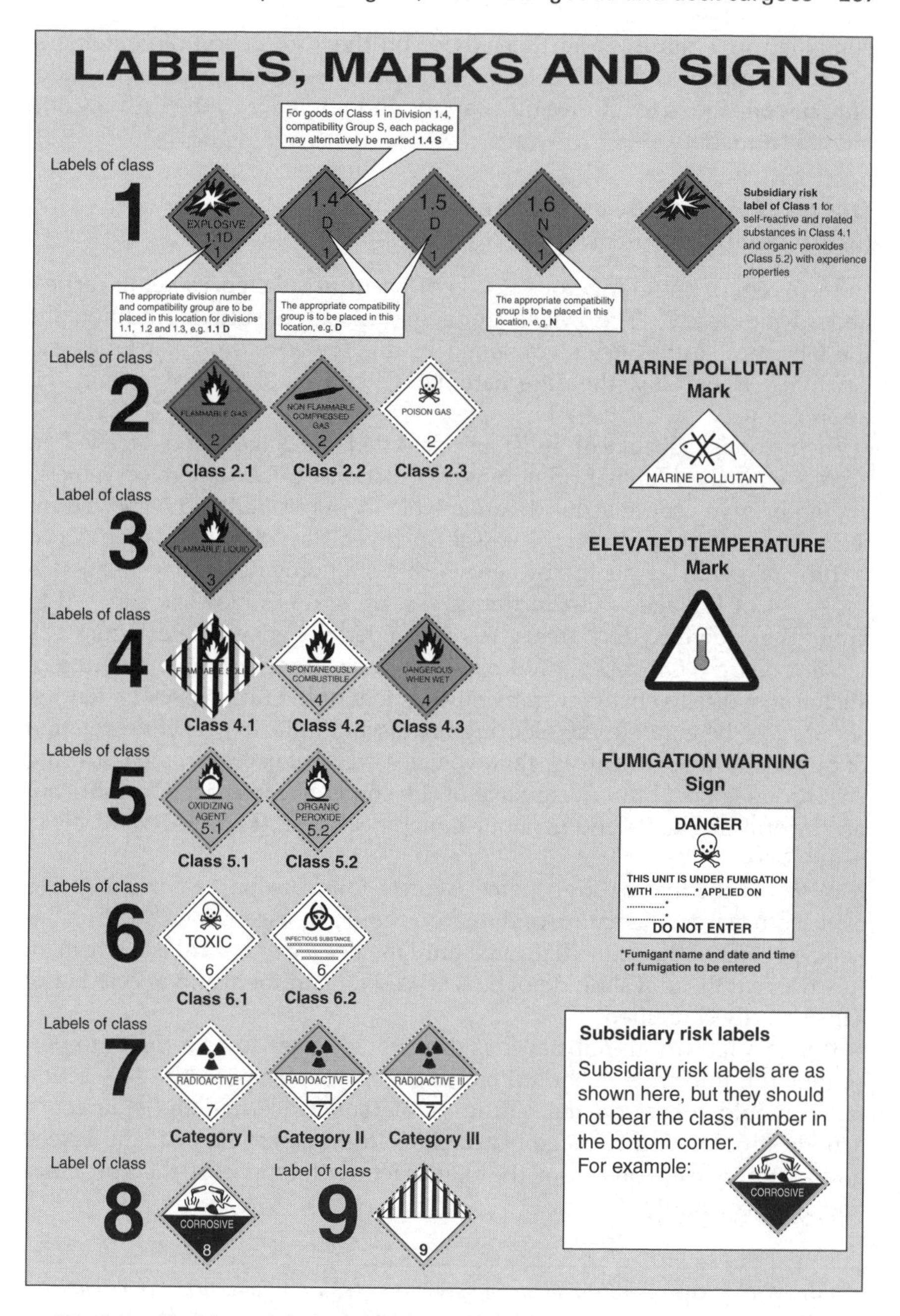

Fig. 9.1 Marking of dangerous classes of goods. Reproduced with kind permission from IMO.

contained in Chapter 7. Further advice on the segregation of containers, housing dangerous goods on board container vessels, is given in the code. Similar conditions for Ro-Ro units apply, and reference to the IMDG Code should be made.

Dangerous/hazardous cargoes

(in dry cargo/container ships or Ro-Ro vessels)

In the event of any dangerous goods or harmful substances being carried aboard the vessel, 'The IMDG Code should be consulted. Additionally, the Chemical Data Sheets contained in the *Tanker Safety Guide (Gas and Chemical)* issued by the International Chamber of Shipping may be appropriate.

Such goods/substances must be classified, packaged and labelled in accord with the Merchant Shipping Regulations. Such trailers or vehicles should be given special consideration when being loaded and inspected for leakage prior to loading on the vessel. Such vehicles/containers should also be provided with adequate stowage which would provide good ventilation in the event of leakage whilst in transit, e.g. upper deck stowage exposed to atmosphere (defined deck area) is recommended as a general rule (Figure 9.2).

Deck (Cargo) Officers should pay particular attention to the securing of such transports to ensure negative movement of the unit. Special attention should also be given to the securing of adjacent units to prevent escalation of cargo shifting in a seaway. Tank vehicles may not necessarily be carrying hazardous goods, but any spillage of the contents could act as a lubricant on surrounding units and generate a major cargo shift on Ro-Ro vessels in heavy seas.

In the event that a cargo parcel/unit is found to be 'leaking' or have exposed hazards, the nature of the cargo should be ascertained and personnel kept clear of the immediate area until the degree of hazard is confirmed. In any event the unit should not be accepted for shipment and rejected until satisfactorily contained.

Where a hazardous substance is discovered at sea to be a threat to personnel, full information should be sought as soon as possible. Any action taken would reflect on the nature of the substance and the emergency actions stipulated in carriage instructions. It may become prudent to seek additional instructions from the manufacturer/shipper of the substance and act accordingly.

> ***Note:*** *With reference to Regulation 54 of SOLAS (1996 Amendment) in ships having Ro-Ro cargo spaces, a separation shall be provided between a closed Ro-Ro cargo space and the weather deck. The separation shall be such as to minimize the passage of dangerous vapours and liquids between such spaces. Alternatively, separation need not be provided if the arrangements of the closed Ro-Ro space are in accordance with those required for the dangerous goods carried on the adjacent weather deck.*

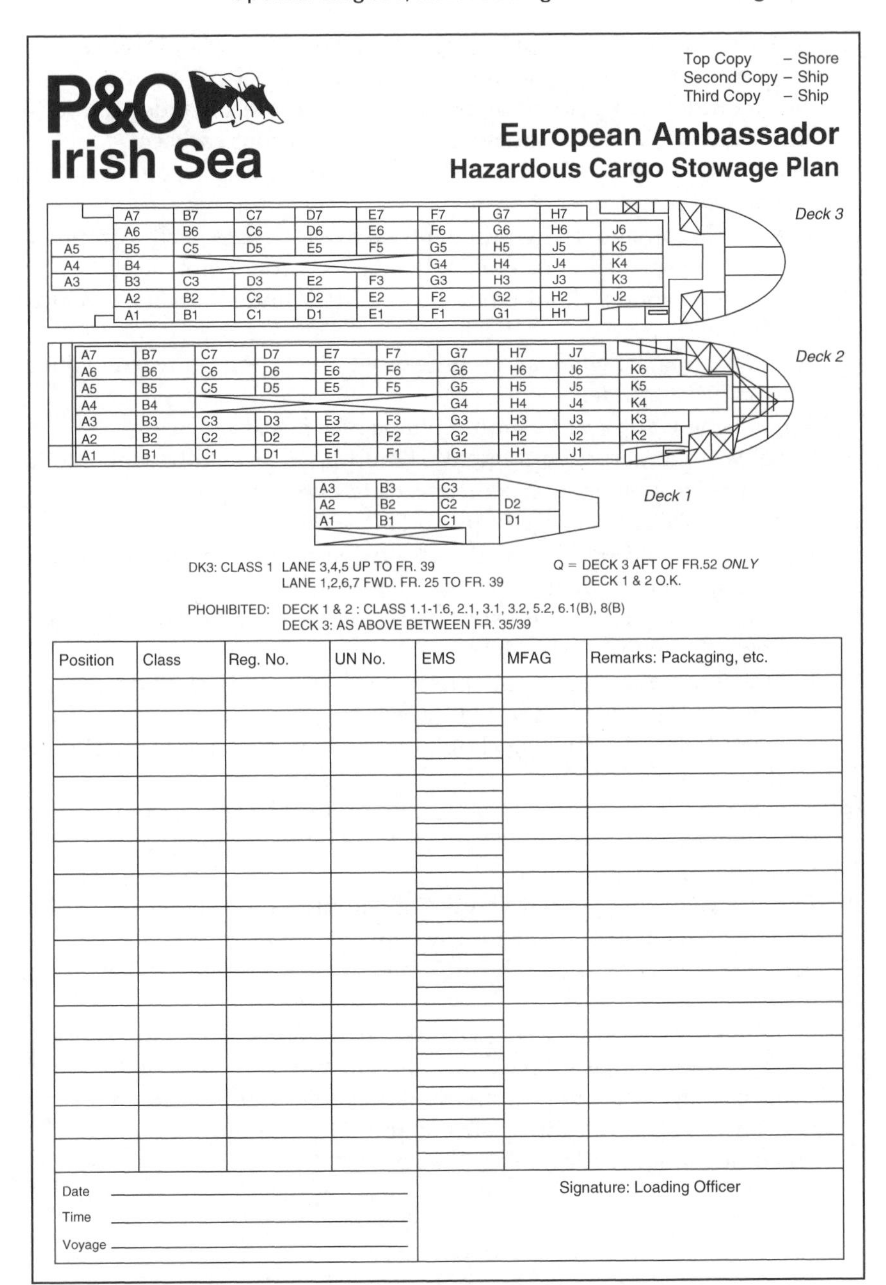

P&O Irish Sea

Top Copy – Shore
Second Copy – Ship
Third Copy – Ship

European Ambassador
Hazardous Cargo Stowage Plan

Deck 3

Deck 2

Deck 1

DK3: CLASS 1 LANE 3,4,5 UP TO FR. 39
LANE 1,2,6,7 FWD. FR. 25 TO FR. 39

Q = DECK 3 AFT OF FR.52 *ONLY*
DECK 1 & 2 O.K.

PHOHIBITED: DECK 1 & 2 : CLASS 1.1-1.6, 2.1, 3.1, 3.2, 5.2, 6.1(B), 8(B)
DECK 3: AS ABOVE BETWEEN FR. 35/39

Position	Class	Reg. No.	UN No.	EMS	MFAG	Remarks: Packaging, etc.

Date ____________
Time ____________
Voyage ____________

Signature: Loading Officer

Fig. 9.2 Stowage and monitoring of hazardous goods in Ro-Ro units.

Precautions when loading/discharging hazardous goods

1. All documentation regarding the 'dangerous goods' should be in order and include the Container Packing Certificate, the Shipper's Declaration and relevant Emergency Information.

2. All cargo operations should be supervised by a responsible officer who will be in possession of operational and emergency information.
3. No unauthorized person, or persons, intoxicated or under the influence of drugs should be allowed near to hazardous cargoes.
4. The compartment or deck area should be dry and clear, suitable for the stowage of the cargo.
5. Where cargo-handling equipment is to be used, such equipment should be inspected to be seen to be in good order before use.
6. Dangerous goods should not be handled under adverse weather conditions.
7. All packaging, labelling and segregation of the goods are carried out as per the IMDG Code.
8. Tanks, where applicable, should not be overfilled.
9. Suitable 'Emergency Equipment' should be kept readily available for any and every potential hazard associated with the goods.
10. Fire wires should be rigged fore and aft of the vessel.
11. Packages should be identified and stowed in an appropriate place to protect against accident.
12. Safe access to packages must be available in order to protect or move away from immediate hazards.
13. Emergency stations with suitable protective clothing should be identified in respect to the location of the cargo.
14. Correct signals, i.e. 'B' Flag should be displayed during the periods of loading and discharging.
15. Stowage positions should be such as to protect the goods from accidental damage due to heating. Combustible materials being stowed away from all sources of ignition.
16. Cargoes requiring special ventilation should be positioned to benefit from the designated ventilation system.
17. The Port Authority should be informed of all movements of hazardous goods.
18. Suitable security should be given to special cargoes like explosives.
19. All hazardous parcels should be tallied in and tallied out of the vessel.
20. Some packages may require daylight movement. Some operations may also be effected by rain or strong sunlight and appropriate loading schedules should reflect related hazards.

Note: *Emergency information on cargoes should include:*

- *The correct technical name of the product and its UN number*
- *Classification and any physical and/or chemical properties*
- *Quantity to load and the designated space to load*
- *The stated action in the event of leakage*
- *Fire fighting and spillage procedures, and any specific equipment required.*

Deck cargoes

The term 'weather deck' refers to an open deck which is exposed to the weather on a minimum of two sides. The phrase is synonymous with deck cargo being carried on exposed decks and running a greater risk of loss because of the stowage location. Below decks and unaffected by the elements of weather provides a level of assumed security completely opposite to the weather risks associated with deck cargoes.

Numerous Codes of Safe Practice, conventions and recommendations have been published to advise on the securing and safe transport of specific deck cargoes, inclusive of timber, containers, vehicles, steelwork, etc. The losses incurred over the years would indicate that the force and power of the elements may generate extreme forces on exposed cargoes, causing restraints to part, and cargo parcels to be lost overside.

Such losses, if noted frequently, would probably deter the carriage of any deck cargoes at all. However, certain cargoes must be categorized and classed as deck cargo because of inherent dangers if they were carried below decks, e.g. certain hazardous goods, such as acids and corrosives.

General observance of SOLAS (Chapters 6 and 7), together with the Code of Safe Working Practice for Cargo Stowage and Securing of Cargo Units Including Containers, have become recognized sources of information. Alongside these, full reference to the ship's Cargo Securing Manual should go some way to assisting the Cargo Officer with decisions concerning the number and positioning of securings and restraints on deck cargo loads.

The regulations require that the following criteria are met when carrying deck cargo:

1. That the vessel will have adequate stability at all stages of the voyage for the amount of cargo it is proposed to load. It should be borne in mind that cargoes like coke and timber may absorb up to about one-third of their own weight by water. Also, losses of bottom weight like fuel, oil and water from double bottom tanks would work against the positive stability of the vessel.
2. Adequate provision must be made for the safe access of the crew when passing from one part of the vessel to another. Deck cargoes that prevent access to crews' quarters, either along or underdeck, must be provided with a walkway over the cargo, and in any event walkways are required for ships with timber deck cargoes.
3. Steering gear arrangements must be protected against damage and in the event of a breakdown of the gear, enough deck space must be available to operate an emergency system.
4. If cargo is to be stowed on hatches, these hatches must be correctly battened down and of adequate strength to support the carriage of the cargoes.
5. Decks, designated for the stowage of deck cargo, must be of adequate strength to support the stowage.

6. Deck cargo parcels are to be well-secured and, if necessary, protected from the weather elements including the heat of the sun. The height of any cargo should not interfere with the navigation of the vessel and obstruct the keeping of an efficient lookout.

Example deck cargoes

Acids and corrosives – liquid acids and dangerous corrosive substances are usually carried in glass containers known as carboys. These containers are straw protected by a steel wire frame and are often crated for shipping. They are always allocated deck stowage away from crews' quarters in accordance with the IMDG Code and would need to be well lashed and secured against movement. In the event of spillage, the accompanying documentation should be consulted and any persons involved in clear up procedures should be issued with protective clothing inclusive of goggles, gloves and suitable footwear.

Chemicals – the type of chemical substance and its form will depend on its style of packaging. Obviously, the numerous chemicals shipped vary considerably and stowage method would be advised by shippers and supplied documentation. Special attention should be paid to instructions in the event of spillage occuring, as some chemicals react with water or air, and become harmful to personnel if incorrect procedures are adopted.

Containers – regularly carried on open decks of container vessels in the 'stack'. However, containers carrying hazardous goods are identified and given appropriate segregated stowage. Where single containers with dangerous goods are carried on open decks on other than dedicated container vessels, suitable stowage and securing are expected to be provided. The main concern for Cargo Officers is that the goods themselves are secured inside the container and packed under correct supervision and delivered for shipment with a Container Packing Certificate, together with relevant documentation regarding the actual goods inside the container.

Gases – carried in cylinders of various sizes. These must be well-secured against unwanted movement. They should not be stowed near any heat source and protected from the sun's rays, usually by a tarpaulin.

Livestock (see Chapter 6) – most livestock would be carried on a sheltered part of the upper decks, along with shipper's instructions for feeding and hygiene.

Oil (drums) – can be carried below decks as well as above decks. Part cargoes are often carried as deck cargo to provide an improved stability condition without having to shut cargo out. Drums are usually of 50 gallon size and should be tightly packed, the most common being for the carriage and shipping of lubricating oil. Once stowed, they should be securely lashed and bowsed into the side bulwarks. If total deck coverage is employed then a walkway, similar to timber deck cargoes, would need to be constructed to provide crew access to fore and aft parts of the ship.

Steelwork – may be shipped in a variety of forms: castings, bulldozers, railway lines, etc. Must be stowed on timber bearers and not steel to steel. The bearers are meant to reduce friction between the deck and load but also spread the deck load capacity weight. In every case, heavy-steel cargoes should be well secured preferably with chains and bottle screws. A combination of chains and wires is also considered as being suitable, depending on the nature of the load. Some loads may lend to being welded to the deck to prevent unwanted movement.

General principles

Deck cargo should be stowed and distributed in a manner that will avoid undue stress on deck areas and ensure that adequate stability is retained throughout the voyage. Certain deck cargoes like timber have the associated danger of absorbing moisture at a position higher than the ship's centre of gravity. With the combined burning off of fuel and the consumption of fresh water from the lower tanks of the vessel the danger to generate a loss of metacentric height (GM) or even create a negative GM is readily apparent. Icing of cargoes, particularly container deck stows, could also be extremely detrimental to the stability of the vessel.

Other cargoes may be large or heavy and generate their own restrictions on the ship. Deck cargoes must not impair the working of the vessel, particularly obstructing the lookouts' duties or preventing access to the working spaces of the vessel. Large cargo parcels could increase the windage experienced by the ship and cause excessive leeway affects and such affects would need to be monitored by Navigation Officers (Figure 9.3).

Fig. 9.3 Securing deck cargoes. Steel pipes seen stowed on upper decks. Chain lashings are stretched across at intermediate lengths and tensioned by ratchet gear once in position. The pipes have been left pre-slung with wire snotters for speed of discharge.

During loading, Chief Officers are advised to ensure that decks are not overstressed by 'point loading' and that supporting structures about the loaded area are adequate to cater for the size and volume of load. All loads must be suitably secured to prevent movement in a seaway, and in the event of heavy weather, prior to sailing.

All deck cargoes should be loaded in accord with the Merchant Shipping (Load Lines) (Deck Cargo) Regulations and S.I. No. 1089 of 1968.

Offshore supply vessels

A major section of the industry is occupied with oil and gas recovery from offshore waters. Offshore installations, from the colossal Production Platforms, to the smaller drilling rigs have the need to be re-supplied on a continuous basis. Cargoes vary in this sector of the industry from the unusual in the form of 'mud' and/or cement, carried in underdeck tanks to the more mundane general stores packed in small containers. The offshore supply vessels, once in close proximity of the installation, are discharged by use of the rig's own cranes. The position of the vessel is held precariously close to the structure of the installation by dynamic position (DP) or by expert ship handling skills of the vessel's Master, weather permitting (Figure 9.4).

Fig. 9.4 Capped drilling pipes seen loaded on the wide beam cargo deck of an offshore supply vessel. The pipes are capped at each end to prevent water retention in the event of the vessel encountering rough weather. Any fluid in quantity being retained amongst deck cargoes could seriously affect the positive stability of the vessel.

Timber

Special regulations apply to the carriage of timber on deck (see Chapter 6). Separate loadlines may apply and specific securing arrangements are recommended as per the Code of Safe Practice for Ships Carrying Timber Deck Cargoes.

Vehicles

It is not unusual to see vehicles carried as deck cargo on board ships other than designated Ro-Ro types of vessel, especially tractors and other farm vehicles. Heavy-lift bulldozers, and similar tracked vehicles are frequently secured on deck or on hatch top squares. Private cars are generally carried below decks as protection from the weather elements is preferred.

Securing of vehicles on deck by means of rope, wire or even chains for heavy plant vehicles is expected for ocean-going vessels. Some form of chocking or tomming may also be desirable. Cargo Officers should pay particular attention to the securing of these cargoes. They are often the last parcels to be loaded and rigging gangs may be tempted to cut corners to be off the ship prior to sailing.

Once at sea, a prudent Chief Officer would order deck cargo lashings to be tightened, especially in the event of a heavy weather warning.

Chapter 10
Security, cargo documentation, stability examples

Introduction

Since the implementation of the International Ship and Port Facility Security (ISPS) Code in July of 2004, all Ship's Officers have been made aware of the need to be security conscious. This is not to say that before this time personnel were ignorant to the dangers and security risks which have always been associated with the maritime industries. The fact that ports have now installed better security fences, X-ray detection methods, close monitoring of dock transports, and tighter control of crews seems to have provided some degree of improved marine security.

For the Cargo Officer, vigilance is essential and on most ship's security starts with ensuring that correct documentation is presented by the crew members on joining, close inspection of the cargo manifests, correct shipping papers for specific cargoes, etc. It is from such information that the safety of the ship can be assured. The Chief Officer is able to take account of the vessel's stability criteria for all stages of the voyage. Hazardous parcels can be secured and monitored for the protection of personnel and cargo alike, while the function of the ship's crew is to protect the shipowner's interests and affect the delivery of all cargoes in good condition and a safe manner.

It has been said that information is power. It is also abundantly clear that cargo information is an essential element of the ship's well-being. To this end the ports around the world are moving rapidly to comply with the security measures required by the Code. Maritime authorities are continuing to work under the umbrella of the International Safety Management (ISM) System and monitoring the safe operation of vessels on the high seas.

Industry sectors such as Safe Navigation operate with external assistance such as VTS schemes, communication networks and hydrographic departments around the globe. The safe transport of cargoes now similarly employs equal support, in the way of customs, police and in some cases the military, to ensure a secure working environment in the modern world. However, these people cannot be all things to all men and it has become clear that the Ship's Officer is closer to the front line of safety and security aboard ships than any other individual.

The ISPS Code and cargo security

Introduction

It is difficult to visualize Cargo Officers being directly involved in the cargo security aspects of an 8000 teu container vessel other than being vigilant during loading, discharging and while in transit. The practicalities of searching excessive numbers of containers are clearly beyond their scope. Sampling possibly a few containers at random must be considered the maximum that anyone could expect as being practical. Security of cargoes must therefore be considered at the start of the container's journey when it is empty, prior to the packing stage. The 'stuffing' of the unit must be carried out under supervision and receive a Packing Certificate. The goods would be subject to customs controls and inspection before being sealed.

Units should be provided with secure holding before delivery to the terminal. Once inside the container park, units fall under the security cordon expected by the ISPS Code. Full documentation of the unit is listed with the shipping agents and seals would be inspected prior to loading the unit on board the vessel.

The ability to detect security threats and take preventive action is paramount. The level of terminal security would vary from port to port and the degree of ship/port interface would be established with experience. It would be envisaged that the Port Security Officer (PSO) would liaise with the Ship Security Officer (SSO) regarding all aspects of 'cargo security'. Such liaison is expected to ensure that:

1. tampering of cargo is prevented, and
2. that cargo which is not intended for shipment, is prevented from being accepted and stored on board.

In order to retain a safe environment it is anticipated that such measures will be in place to include inventory and control applications, such precautions being supported by the identification of all cargo parcels on board the vessel. To this end, container companies have installed methods that allow the tracking of all 'box' units and Roll-on, Roll-off (Ro-Ro) units, showing as being approved for loading and shipping by the vessel.

Screening of stores, cargo parcels and unaccompanied baggage tends to rest with the port facility and is meant to be covered by the 'Port Facility Security Plan' (PFSP). Such screening may include the searching of baggage both ashore and on board. Scanning equipment and/or specially trained dogs may very well be used, to ensure the security of packages.

The ISPS Code

Application

The ISPS Code is applicable to the following types of vessel on international voyages:

1. Passenger ships, inclusive of high-speed passenger craft
2. Cargo ships, including high-speed craft of 500 gross tonnage and upwards
3. Mobile offshore drilling units
4. Port facilities serving such ships engaged on international voyages.

Definitions effective within the ISPS Code

Ship Security Plan – means a plan developed to ensure the application of measures on board the ship designed to protect persons on board, cargo, cargo transport units, ship's stores, or the ship, from risks of a security incident.

Port Security Plan – means a plan developed to ensure the application of measures designed to protect the port facility and ships, persons, cargo, cargo transport units, and ships stores within the port facility from the risks of a security incident.

Ship Security Officer – that person on board the ship accountable to the master, designated by the company as responsible for the security of the ship, including implementation and maintenance of the Ship Security Plan and for the liaison with Port Facility Security Officers.
Note: The master can now be the designated SSO.

Company Security Officer – means that person designated by the company for ensuring that a ship security assessment is carried out; that a Ship's Security Plan is developed, submitted for approval and thereafter implemented and maintained, and for liaison with Port Facility Security Officers and the SSO.

Port Facility Security Officer – means the person designated as responsible for the development, implementation, revision and maintenance of the PFSP and for liaison with the SSO and the Company Security Officer (CSO).

Security 'Level 1' – means that level for which minimum appropriate protective security measures shall be maintained at all times.

Security 'Level 2' – means that level for which appropriate additional protective security measures shall be maintained for a period of time as a result of heightened risk of a security incident.

Security 'Level 3' – means that level for which further specific protective security measures shall be maintained for a limited period of time when a security incident is probable or imminent, although it may not be possible to identify the specific target.

Ship – the term ship as used within the context of the code includes mobile offshore drilling units and high-speed craft as defined by Regulation XI-2/1.

Ship/port interface – means the interactions that occur when a ship is directly and immediately affected by actions involving the movement of persons, goods or the provisions of port services to or from the ship.

Ship-to-ship activity – means any activity not related to a port facility that involves the transfer of goods or persons from one ship to another.

Security incident – means any suspicious act or circumstance threatening the security of the ship, or of a port facility or any ship/port interface or ship-to-ship activity.

Security level – means the qualification of the degree of risk that a security incident will be attempted or will occur.

Security threats

Clearly in this day and age any threat to the ship or the port's facilities could have a direct consequence to personnel working aboard or within the port confines. The PFSP is meant to identify such threats and prioritize protective security actions. Such threats may take on a variety of forms from damage to the ship or port facilities – i.e. from an explosive device, arson, tampering with cargo, smuggling activities – to the extreme of nuclear, biological or chemical attack.

Cargo concerns

Anything that generates cause to affect the well-being of cargo parcels is of concern to Ship's Officers. Pilferage from open stow cargoes has long been an expensive activity and to some extent containerization went a long way to curb theft. However, thieves are known to hijack the whole container, often with insider knowledge as to its contents. Other cargoes are of a higher profile and more readily visible, requiring immediate and increased security – like, for instance, nuclear waste flasks (Figures 10.1 and 10.2).

Shipboard security activity

Many aspects of shipboard activity are exposed to abuse and threats to security. Some of these are listed and officers should be mindful of the security elements associated with the following:

1. The handling of cargo which may contain harmful substances or terrorist personnel
2. The handling of unaccompanied baggage
3. The handling and loading of ship's stores
4. Controlling access of persons who may have criminal intent
5. Monitoring berthing areas in close proximity to the ship's hull
6. Monitoring offshore areas to prevent incursion by waterborne craft

Fig. 10.1 Nuclear waste flasks discharged with full heavy-lift precautions onto special transports in Japan monitored by security mobiles and specialized personnel. Open aspects and security fencing surround the working area of the port facility.

Fig. 10.2 Security personnel monitor nuclear waste flasks, following discharge into the port facility handling grounds. Customized transports are designated to each flask and movement is checked at every stage of transit.

7. Controlling the embarkation of persons and their effects (especially so with high-profile vessels carrying increased numbers of passengers) (Figure 10.3).

Fig. 10.3 High-profile passenger ship 'Queen Mary 2' lies port side to alongside the berth in Southampton. Such vessels require maximum security within the port facility and on all access points to the vessel.

Security progress

In order to be compliant with the ISPS code, Safety of Life at Sea (SOLAS) has been amended to include relevant requirements:

1. All ships on international voyages will be equipped with Automatic Identification System (AIS).
2. Companies will be expected to install Ship Security Alert Systems on their vessels.
3. Ships will run a continuous synopsis record while in service from the time of launching for new builds.
4. Companies must report to a flag, state-appointed, recognized security organization and create internal positions for dedicated CSO as well as on board SSO.
5. Companies will develop their own security plans to suit each individual vessel, which must be implemented on board.
6. Vessels which are compliant will be issued and International Ship Security Certificate (ISSC) by the companies recognized safety organization. This organization will also be responsible for conducting internal audits.

Application

The ramifications of the ISPS Code will mean that practical activities to safeguard the ship and cargoes will begin to operate; for example, masters being informed beforehand of the presence of self-igniting chemicals; containers being scanned inside the terminal before being loaded aboard the vessel; while tighter access controls into terminals and onto ships will expect to create a more security conscious environment.

Cargo documentation summary

Ballast Management Record – the dangers to aquatic life have instigated the need for tighter controls on the movement and in particular, the discharge of ballast waters. Positions of ballast change, dates, amount and tank location.

Bill of Lading (B/L) – are the consignee's title to the goods which have been shipped or are about to be shipped. The B/L will quantify the goods and refer their condition at the time of shipping and he/she would expect to receive the goods at the port of discharge in the same good condition as when shipped. In the event that the goods are damaged at receipt or in loading or discharging the B/L would be endorsed to specify the damage. Such an endorsed B/L would be considered as a dirty or foul B/L, as opposed to a clean B/L which is without endorsement. The B/Ls are usually drawn up by the shipping agent and signed by the Master of the Vessel.

Cargo manifest – the official listing of all cargo parcels carried on board the vessel. This document is what the master bases his declaration on when entering port; all cargoes being officially declared on the manifest which is subject to inspection by Customs Officers, and port security inspection.

Cargo Record Book – vessel engaged in the carriage of noxious liquid substances must carry a record of the cargo movements affecting the ship. The same ships would also be expected to carry a Maritime and Coastguard Agency (MCA) approved *Procedures and Arrangement Manual*, reflecting the operational aspects of the vessel.

Cargo Securing Manual – a legal requirement for every ship other than those engaged in the carriage of solid or liquid bulk cargoes. The purpose of the manual is to cover all relevant aspects of cargo stowage and securing. Securing devices and methods must meet acceptable criteria for strength, applicable to relevant cargo units, inclusive of containers and Ro-Ro transports. Each manual is prepared in a manner to reflect the individual ship's needs, relevant to the type of cargo parcels it is engaged to ship.

Cargo stowage plan – a charted plan of the vessel's cargo-carrying spaces which illustrates the type, tonnage and description of goods for designated discharge in the various intended ports of call for the voyage. The plan is constructed by the Cargo Officer and is meant to provide an overall illustration of the distribution of the ship's cargo. The plan is copied and despatched to the various ports of discharge prior to the ship's arrival on the berth. It allows relevant cranes to be ordered and stevedore gangs to be employed in advance which subsequently speeds up the time of the vessel lying in port. It is considered essential for dry cargo vessels, tankers, bulk carriers and container vessels to all carry stowage plans respective to their relevant cargoes.

Certificate of Fitness – is required by every UK tanker and gas carrier. These are issued by the MCA and are valid for a period not exceeding five (5) years, being subject to initial, annual and intermediate surveys. This certificate cannot be extended.

Charter party – is a private contract between the principal parties to an agreement and is evidence of who the operator of the ship is. Charter parties are set in three categories: Time Charter, Voyage Charter or a Demise Charter (US Bare Boat Charter). Variations of the three categories are drawn up based on the operational requirements of the ship and the intended voyage.

Container Packing Certificate – the packing and unpacking of containers usually takes place at shore terminals or at the address of the shipper or consignee. Prior to loading on board the vessel, a valid Container Packing Certificate must be received as evidence that the goods have been packed in such a way as to withstand carriage at sea. It is also a security check that the unit does not contain contraband goods and that the merchandise is as what is described on the certificate. Container units are now electronically scanned at entry to many shipping terminals.

Document of Authorization – is a required certificate issued by a surveyor following survey of the ship's cargo holds and its ability to carry cargo safely. Unless the ship is in the possession of an Exemption Certificate, the Document of Authorization would be an official requirement.

Document of Compliance (dangerous goods) – a certificate of compliance issued to a type of ship that is permitted to carry certain categories of hazardous/dangerous goods. Not all ships can carry hazardous cargoes; for example, passenger vessels are not allowed to carry Class 1, explosives.

Enclosed space entry permit – a work permit which is issued prior to entry into an enclosed space compartment. The permit is issued only after all the required safety checks and inspections of the compartment have been made and the relevant precautions have been taken.

Export licences – these are supplied by the shipper as required for certain specific cargoes: computers, foodstuffs, livestock, armaments, etc. The export licence is required by Government/State Officials for certain types of cargoes which are subject to inspection by customs (e.g. armaments, drugs, etc.).

International Security Certificate – issued to a vessel by a recognized security organization confirming that the ship is compliant to the ISPS Code.

Mate's Receipt – a receipt for goods received and delivered on board the vessel. As the name implies, it is signed and issued by the Mate of the Ship, i.e. the Chief Officer. It may form the basis for the final B/L.

Note of Protest – is where the Master of a Ship makes a declaration of 'Protest' under oath before a Notary Public, Magistrate or British Consul. The declaration often affects cargo damaged or suspected of having damage due to a 'peril of the sea'. The main use of Protest in the UK is to support a cargo owner's claim against his underwriters. The Note of Protest is admissible as evidence before legal tribunals in many countries, but not in the UK unless both parties agree. Masters should note Protest as soon after arrival in port and before 'breaking bulk'. The master may extend 'Protest' once the situation has been further assessed and the full extent of damage is revealed.

Register of Lifting Appliances and Cargo-Handling Gear – a record of all the ship's cargo-handling equipment, usually retained and updated by the Ship's Chief Officer. It contains all the certificates for such items as shackles, blocks, wires, derrick and crane tests, hooks, chains, etc. The register is open to inspection by Port State Control Officers and would be required by the surveyor when carrying out the Cargo-Handling Equipment Survey.

Rigging plan – a ship's arrangement plan which illustrates the operational aspects of the ship's lifting appliances. Safe working loads and maximum permissible outreach limits would expect to be displayed alongside the related positions of cargo stowage compartments.

Stability information booklet – the ship's stability criteria may be in booklet format or in the form of a series of plans, or even carried in a combination format of both. Either way the documents are in the control of the Ship's Chief Officer and will include the following: general particulars of the vessel; a general arrangement plan showing cargo compartments and tank dispositions; special notes on the stability and loading procedures; hydrostatic particulars; metric conversion table; capacity plan showing centre of gravity of cargo stowage compartments (to include free surface moment of oil and water tanks); notes on the use of free surface moments; cross curves of stability (known as KN curves) with examples of their use; deadweight scale; list of ship conditions and typical condition sheets; statical stability curve for conditions; simplified stability information together with damaged stability criteria.

Transportable Moisture Limit (TML) Certificate – a certificate issued within 7 days of measuring the moisture limit of the bulk product to be shipped.

Abbreviations used in the following example calculations

A	Aft
AP	Aft perpendicular
C of G	Centre of gravity
cm	Centimetre
CoT	Change of trim
d	distance
DW	Dock water
DWA	Dock water allowance
F	Forward
FP	Forward perpendicular
FWA	Fresh water allowance

G	Position of the ship's C of G
GG_1	Movement distance of the ship's C of G
GM	Metacentric height
KG	Measured distance between the keel and the C of G of the vessel
KM	Measured distance between the keel and the metacentre
L	Length of ship
l	A proportionate length of the ship's length
M	Metacentre
m	metres
MCTC	Moment to change trim 1 cm
mm	Millimetres
RD	Relative density
Stbd	Starboard
SW	Salt water
tan	tangent
TPC	Tonnes per centimetre
W	Displacement of vessel
w	added or discharged weight

Cargo work – stability examples

Example 1

A vessel of 5870 tonnes displacement has a load draught of 5.4 m with a TPC = 11. Calculate the load draught of the vessel if she is a working cargo in fresh water.

$$\text{FWA} = \frac{W}{4 \times \text{TPC}} = \frac{5870}{4 \times 11} = 133.4 \text{ mm}$$

$$= 0.133 \text{ m}$$

$$\text{Load draught in FW} = 5.40 + 0.133 = 5.533 \text{ m.}$$

Example 2

A vessel has a load draught in SW of 6.4 m. Calculate the maximum load draught in DW of RD 1.010. The ship's FWA is 75 mm.

$$\text{DWA} = \text{FWA} \times \frac{1025 - \text{density of DW}}{25}$$

$$= 75 \times \frac{1025 - 1010}{25}$$

$$= 75 \times \frac{15}{25}$$

$$= 45 \text{ mm} = 0.045 \text{ m}$$

$$\text{Maximum draught} = 6.40 + 0.045 = 6.445 \text{ m.}$$

Note: *DWA being the amount the vessel may legally submerge her disc (Plimsoll Line) when loading in a DW of less density than that of sea water.*

Example 3

A vessel of 10 000 tonnes displacement with a KG of 7.0 m loads 100 tonnes of KG 12 m. Calculate the new KG of the vessel, by taking moments about the keel.

Weight	*KG*	*Moment*
10 000	7.0	70 000
+100	12.0	+1200
10 100		71 200

$$\text{Final KG} = \frac{\text{total moment}}{\text{total weight}} = \frac{71\,200}{10\,100}$$
$$= 7.0495 \text{ m.}$$

Example 4

A vessel of 12 000 tonnes displacement has a KG of 7.8 m and a KM of 8.6 m. She then loads the following cargo parcels:

250 tonnes at KG of 11.0 m
100 tonnes at KG of 7.0 m and
50 tonnes at KG of 3.0 m

Calculate the vessel's final GM after completion of loading.

Weight	*KG*	*Moment*
12 000	7.8	93 600
+250	11.0	+2750
+100	7.0	+700
+50	3.0	+150
12 400		97 200

$$\text{Final KG} = \frac{\text{total moment}}{\text{total weight}} = \frac{97\,200}{12\,400} = 7.839 \text{ m.}$$

$$\text{GM} = \text{KM} - \text{KG} = 8.6 - 7.839$$
$$= 0.761 \text{ m.}$$

Example 5

A vessel of 7500 tonnes displacement with KG of 6.0 m, and KM of 6.8 m, is expected to load timber on deck in a position of KG 11.0 m. Calculate the maximum weight of timber that can be loaded in order to arrive at the destination with a GM of 0.2 m if an allowance of 15% increase in weight is anticipated with water absorption by the deck cargo.

Let the weight of cargo to be loaded = w

Weight	*KG*	*Moment*
7500	6.0	45 000
1.15w	11.0	12.65w
7500 + 1.15w		45 000 + 12.65w

$$\text{Final KG} = 6.8 - 0.2 = 6.6$$

But

$$\text{Final KG} = \frac{\text{total moment}}{\text{total weight}} = \frac{45\,000 + 12.65w}{7500 + 1.15w}$$

$$\therefore 6.6 = \frac{45\,000 + 12.65w}{7500 + 1.15w}$$

$$\therefore 49\,500 + 7.59w = 45\,000 + 12.65w$$

$$4500 = 5.06w$$

$$w = \frac{4500}{5.06}$$

$w = 889$ tonnes of timber to load.

Example 6

A ship of 10 000 tonnes displacement is to load a heavy lift of 100 tonnes with a KG of 3.0 m by means of the ship's heavy derrick. The head of the derrick is 24 m above the keel. The ship's KM was 7.0 m, with a KG of 6.2 m before loading. The load is to be stowed on the ship at a KG of 6.0 m.

Calculate: (a) the minimum GM experienced and (b) the final GM.

(a)

$$GG_1 = \frac{w \times d}{W}$$

$$= \frac{100 \times (24 - 3)}{10\,000}$$

$$= 0.21 \text{ m}$$

$$\therefore KG = 6.2 + 0.21 = 6.41 \text{ m}$$

$$\therefore GM = 7.0 - 6.41 = 0.59 \text{ m.}$$

(b)

$$GG_1 = \frac{100 \times (6 - 3)}{10\,000} = 0.03\,\text{m}$$

$$\therefore \text{KG} = 6.2 + 0.03 = 6.23\,\text{m}$$
$$\therefore \text{GM} = 7.0 - 6.23 = 0.77\,\text{m}.$$

Example 7

Note: When the C of G of a vessel moves off centre, an upsetting lever is produced which causes the vessel to list until G and M are in the same vertical line. The angle of heel due to G being off centre is found by the formula

$$\tan\theta = \frac{\text{Transverse } GG_1}{\text{GM}} \quad \text{but} \quad GG_1 = \frac{w \times d}{W}$$

$$\therefore \tan\theta = \frac{w \times d}{W \times \text{GM}} = \frac{\text{listing moment}}{W \times \text{GM}}$$

A vessel with 4000 tonnes displacement which is initially upright moves a 12 tonnes weight 7 m transversely across the deck. The ship's GM with the weight on board is 0.3 m. Calculate the resulting list.

$$\tan\theta = \frac{w \times d}{W \times \text{GM}} = \frac{12 \times 7}{4000 \times 0.3}$$
$$\theta = 4°0'.$$

Example 8

A vessel of 11 000 tonnes initial displacement loads a 50 tonnes weight by a floating crane in a position 12 m to port off the ship's centre line. Assume that the KG and KM remain constant and that the vessel is upright prior to loading. Calculate the angle of list if the ship's GM is currently 0.25 m.

$$\tan\theta = \frac{w \times d}{W \times \text{GM}} = \frac{50 \times 12}{11050 \times 0.25}$$
$$\therefore \theta = 12°25' \text{ to port.}$$

Example 9

A vessel of 10 000 tonnes displacement with an initial GM of 0.2 m conducts the following cargo operations:

Loads 50 tonnes 4 m to Stbd of the centreline.
Loads 70 tonnes 5 m to port of the centreline.
Discharges 90 tonnes 3 m to Stbd of the centreline.
Shifts 40 tonnes 6 m to Stbd.
Assuming KG and KM remain constant, determine the final list.

Weight	*Distance off centre*	*List moment*	
		Port	*Stbd*
10 000	–	–	–
+50	4 m Stbd		200
+70	5 m port	350	
−90 (discharge)	3 m Stbd	270	
+40 (transferred)	6 m Stbd		240
10 030		620	440
		440	
		180 (port)	

$$\therefore \tan\theta = \frac{180}{10\,030 \times 0.2}$$

$$\theta = 5°13' \text{ to port.}$$

Example 10

A vessel of 160 m length, loads 40 tonnes in a position 60 m from the AP. Calculate the final draughts if the initial draughts are forward 5.0 m aft 6.0 m. The longitude centre of floatation is 70 m forward of the AP. The ships TPC is 20, and MCTC is 100.

$$\text{Sinkage due to load} = \frac{\text{weight loaded}}{\text{TPC}} = \frac{40}{20} = 2 \text{ cm}$$

$$\text{CoT} = \frac{w \times d}{\text{MCTC}} = \frac{40 \times 10}{100} = 4 \text{ cm (by stern)}$$

$$\text{CoT aft due to CoT} = \frac{l}{L} \times \text{CoT} = \frac{70}{160} \times 4 = 1.75 \text{ cm}$$

$$\text{CoT forward due to CoT} = 4 - 1.75 = 2.25 \text{ cm}$$

	Forward	*Aft*
Initial draughts	5.0	6.0
Sinkage	0.02	0.02
	5.02	6.02
CoT	−0.0225	+0.0175
Final draughts	4.9975F	6.0375A.

Appendix A
Miscellaneous cargo information

Introduction

A text of this size cannot hope to cover every commodity or every situation that Ship's Officers encounter. Cargoes are varied and can be carried in many alternative forms. The following brief notes are meant to portray the fundamentals that go along with and support the various chapters and specifics expressed throughout this work.

Charter Party

A formal agreement to hire, rent or lease a ship. It is a private contract between two parties and may be written up in several formats, i.e. Time Charter, Bare Boat Charter or a Demise Charter.

Demurrage

Monies paid by the Charterer to the owner of the ship for delays in loading and unloading in accord with the terms of the Charter Party. The rate of demurrage is fixed and agreed between the owner and Charterer at the time of completing the Charter Party and cannot be altered. Can be taken in contrast with 'Despatch Money' opposite to demurrage where the owner pays reward money to the Charterer for completing loading and/or discharging earlier than the expiry of 'Laytime'.

Freeboard Form (FRE 13)

Following the completion of cargo operations in a port, the ship's draughts will expect to have changed. The draughts fore and aft should be read and the mean draught obtained.

It is a legal requirement that before the ship sails outward, the 'Freeboard Certificate' is completed and displayed for the crew to see that the vessel is not overloaded and the minimum bow height is not exceeded (Figure A.1).

DRAUGHT OF WATER AND FREEBOARD NOTICE

For

Issued by the Department of Transport pursuant to Section 10 (2) of the Merchant Shipping (Load Lines) Act 1967

SHIP ....PORT OF REGISTRY ....

GROSS TONNAGE ....

(Where a ship has alternative gross tonnages, both must be given)

(1) Summer freeboard* .... millimetres corresponding to a mean draught † of .... millimetres (equal to .... feet .... inches).
(2) Winter freeboard* .... millimetres corresponding to a mean draught † of .... millimetres (equal to .... feet .... inches).
(3) Tropical freeboard* .... millimetres corresponding to a mean draught † of .... millimetres (equal to .... feet .... inches).
(4) Winter North Atlantic freeboard* .... millimetres corresponding to a mean draught † of .... millimetres (equal to .... feet .... inches).
(5) Allowance for fresh water for all freeboards other than timber freeboards* .... millimetres.
(6) Timber Summer freeboard* .... millimetres corresponding to a mean draught † of .... millimetres (equal to .... feet .... inches).
(7) Timber Winter freeboard* .... millimetres corresponding to a mean draught † of .... millimetres (equal to .... feet .... inches).
(8) Timber Tropical freeboard* .... millimetres corresponding to a mean draught † of .... millimetres (equal to............ feet .... inches).
(9) Timber Winter North Atlantic freeboard* .... millimetres corresponding to a mean draught † of .... millimetres (equal to .... feet .... inches).
(10) Allowance for fresh water for timber freeboards* .... millimetres.

* Particulars to be given above of freeboards and allowance for fresh water are to be taken from the load line certificate currently in force in respect of the ship. Paragraphs referring to freeboards which the certificate shows have not been assigned to the ship must must be deleted.

† The mean draught to be given below is the mean of the draughts which would be shown on the scales of measurement on the stem and on the stem post of the ship if it were so loaded that the upper edge of the load line on each side of the ship appropriate to the particular freeboard were on the surface of the water.

Where the draught is shown on the scales of measurement on the stem and on the stem post of the ship in feet the mean draught must be given in both millimetres and feet and inches using an equivalent of 25.4 millimetres to one inch.

PARTICULARS OF LOADING

1	2	3	4	5	6	7	8	9
Date	Place	ACTUAL DRAUGHT			MEAN FREEBOARD		SIGNATURE OF MASTER AND AN OFFICER	
		Forward	Aft	Mean	Actual (see notes 1 and 2 below)	Corrected (see note 3 below)	Master	An Officer

NOTES (1) The actual mean freeboard (column 6) is the mean of the freeboards on each side of the ship at the time when the ship is loaded and ready to leave.

(2) If the actual mean freeboard is less than the appropriate minimum salt water freeboard as shown on the load line certificate there must be entered in Column 7 the corrected freeboard arrived at after making any allowances for density of water, rubbish to be discharged overboard and fuel, water and stores to be consumed on any stretch of river or inland water, being allowances duly entered in the ship's official log book.

(3) If the actual mean freeboard is greater than the appropriate salt water freeboard, Column 7 need not be filled in.

This Notice should be posted in some conspicuous place on board the ship, where it can be seen by all members of the crew, before the ship leaves any dock, wharf, harbour or other place for the purpose of proceeding to sea and is to be kept so posted until after the ship arrives at any other dock, wharf, harbour or place.

The date and time of recording the above particulars on each occasion must be entered in the Official Log Book.

This Form should not be handed in with the Official Log Book and Agreement at the termination of the Agreement. It can be used until completed, when a further copy should be obtained.

Fig. A.1.

Freight

Charges made for carriage of the cargo. Advance freight is money paid before the delivery of the cargo – freight being payable concurrently with delivery of the cargo at the port of destination.

Heavy weather and cargo procedures

In the event of heavy weather possibly affecting a ship's passage, certain obvious precautions, depending on the nature of the cargo carried, can be adopted to protect the cargo condition:

1. Investigate an alternative route for the vessel clear of weather-affected areas.
2. Improve the ship's stability and reduce any free surface effects.
3. Tighten up on any cargo lashings, especially deck cargo lashings and heavy lifts.
4. Reduce speed in ample time to avoid the vessel pounding.
5. Adjust the ship's head to avoid excessive rolling.
6. Close up ventilators to avoid water ingress.
7. Check all hatch and access seals are secure.

Laytime

Described as that time which is available to the Charterer to load or discharge the Chartered Cargo, free of any charge other than the freight charges. Laytime cannot commence until three conditions have been satisfied that:

1. the vessel is an arrived ship (under legal terms)
2. Notice of Arrival has been tended (it does not need to have been accepted)
3. the vessel is in all respects ready to load or discharge.

Lifting plant – tests and examination

It is a legal requirement that lifting apparatus like derricks and cranes are periodically tested:

- after installation when new
- following any defect and major repair to the plant
- at intervals not exceeding five (5) years.

Additionally, it would be considered a normal practice for the Cargo Officer to inspect the lifting appliances prior to their use. Each apparatus would also be thoroughly examined by a responsible person (i.e. the Chief Officer) at least every twelve (12) months.

Notice of protest

Masters would 'Note Protest' for any of the following cargo reasons:

1. Whether damage has been caused or is suspected of being caused to cargo.
2. Whenever the ship has encountered heavy weather which may have resulted in damage to cargo.

3. Where cargo is known to have been damaged through a peril of the sea.
4. Where cargo is shipped in such a state that it is likely to deteriorate during the voyage (Bills of Lading (B/L) should be also endorsed in this case).
5. Where a serious breach of the Charter Party has occurred by the Charterer or his agent, e.g. refusing to load cargo, or delaying the loading, loading improper cargo.
6. In all cases of general average being declared.
7. When consignees fail to take delivery of cargo and pay due freight in accord with the terms of the Charter Party or B/L.

Oil-filtering and monitoring equipment (oil water separators)

Every ship of 400 GT and above, but less than 10 000 GT, shall be fitted with oil-filtering equipment which complies with an approved design in accordance with the specifications for such equipment as set out in the recommendations on International Performance and Test Specifications for oily water separating equipment and oil content meters.

Every ship over 10 000 GT and above shall be provided with:

1. oil-filtering equipment (as stated)
2. oil-content-measuring equipment fitted with a 15 parts per million (ppm) alarm device and with arrangements for automatically stopping any discharge of oily mixture when the oil content in the effluent exceeds 15 ppm.

Oil record books

Current legislation require oil tanker vessels to carry two (2) Oil Record Books, one for cargo movement and one for fuel movements. Non-tanker type vessels are only required to carry one (1) Oil Record Book.

Entries into Oil Record Books should cover any movement of oil in or out of the vessel, including internal transfers between tanks. Each entry in the book should be signed by the Ship's Master and another officer.

Security

The Master of the Vessel should not be constrained by the company, the Charterer or any other person, from making a decision which, in his professional judgement as the master, is necessary to maintain the safety and security of the ship. This includes the loading of cargo, including containers or other enclosed cargo transport units, which may have a direct threat to the well-being of the vessel or personnel on-board.

Vapour-recovery systems (protecting the environment)

More and more use is being made of Vapour-Recovery Systems in order to reduce emissions of volatile organic compounds (VOCs) in offshore operations. This expansion is currently being pushed by Norway, which is introducing legislation to ensure that all offshore loading operations have reduction plants in operation by 2005. When cargo tanks are loaded with crude, hydrocarbon gas emissions are vented to air and the loss may be estimated to between 100 and 300 tonnes for each loading. The recovery systems are meant to be economic and currently three types are employed:

1. *Absorption* of non-methane VOCs (NMVOCs) into the crude being loaded.
2. *Liquefaction and the storage* of NMVOCs to be discharged at a later time, or used as onboard fuel.
3. *Sequential transfers* of tank atmospheres during the cargo operations of loading and discharging.

Reliquefied VOC emissions can be used to fuel the VOC recovery plant as well as similar steam-driven on-board systems. It can also be used as a blanket gas in cargo tanks instead of inert combustion gases. Further use can be achieved by discharging it ashore for use as fuel or further refining.

Appendix B

Self-examiner – questions and recommended answers to cargo related examples

Calculations and questions

Example 1

Calculate by how many millimetres a ship may submerge her loadline when she is currently loading in dock water of relative density (RD) 1.013, if the vessel has a fresh water allowance (FWA) of 190 mm.

$$\begin{aligned}\text{Dock Water Allowance} &= \text{FWA} \times \frac{(1.025 - \text{water density number})}{1.025 - 1.000}\\ &= 190 \times \frac{(1.025 - 1.013)}{0.025}\\ &= 190 \times \frac{(0.012)}{0.025}\\ &= 91\ \text{mm}\end{aligned}$$

Summer loadline may be submerged by 91 mm.

Example 2

A rectangular tank of 9 m length and 6 m breadth has a depth from the ullage plug of 11 m. Find how many tonnes of oil of a RD of 0.83 does the tank contain when the ullage is 350 mm?

$$\begin{aligned}\text{Depth of tank} &= 11.0\ \text{m}\\ \text{Ullage} &= \underline{0.35\ \text{m}}\\ \text{Depth of oil} &= 10.65\ \text{m}\end{aligned}$$

$$\begin{aligned}\text{Therefore the volume of oil} &= 9 \times 6 \times 10.65\ \text{m}^3\\ &= 575.1\ \text{m}^3\\ \text{Weight of Oil} &= \text{Volume} \times \text{Density}\\ &= 575.1 \times 0.83\\ &= 477.33\ \text{tonnes.}\end{aligned}$$

Example 3

What is the smallest purchase that could be used to lift a 5 tonne weight with flexible steel wire rope having a safe working load (SWL) of the wire equal to 3.125 tonnes?

$$S \times P = W + \frac{nW}{10}$$

Assuming the purchase is to be used to disadvantage P = n.

$$\text{Then} \quad 3.125 \times P = \frac{(5 \times P)}{10} + 5$$
$$2.625P = 5$$
$$P = 2$$

Therefore a gun tackle is the minimum purchase to use for this lift.

Example 4

Calculate the maximum number of tonnes which can still be loaded into a vessel whose tonnes per centimetre (TPC) = 19 and FWA = 190 mm. Her loaded salt water freeboard is 2310 mm and her present freeboards are 2420 (starboard (Stbd)) and 2404 mm (port), in water of RD 1.009.

Present freeboards 2420 mm Stbd, 2404 mm Port } true mean freeboard = 2412 mm

$$DWA = 190 \times \frac{(1.025 - 1.009)}{1.025 - 1.000}$$
$$= 121.6 \text{ mm}$$

Corresponding salt freeboard

2412 + 121.6 = 2533.6 mm
Permitted freeboard = 2310 mm
Sinkage allowed = 223.6 mm
(TPC = 19, TPmm = 1.9) × 1.9
Cargo to load = 424.84 tonnes

where TPmm represents tonnes per millimetre.

Example 5

A cargo tank with an area of 75 m^2 is being filled from a pipe of 200 mm in diameter. The ullage is now 1.6 m. Calculate how much longer the filling valve must be left open to obtain an ullage of 800 mm, if the average rate of flow through the pipe is 1.75 m/s?

Difference in ullages $= 1600\text{ mm} - 800\text{ mm} = 800\text{ mm (0.8 m)}$

Volume of liquid to load $= 75\text{ m}^2 \times 0.8\text{ m} = 60\text{ m}^3$

Area of pipe $= \pi r^2 = 3.1416 \times 100 \times 100$

$= 31416\text{ mm}^2$

$$= \frac{31416}{1000 \times 1000}\text{ m}^2$$

Volume of liquid loaded per second

$$= \frac{31416}{1000 \times 1000} \times 1.75\text{ m}^3$$

Time to load 60 m³ $$= \frac{60 \times 1000 \times 1000}{31416 \times 1.75}\text{ s}$$

$= 1091\text{ s or } 18.18\text{ m}$

The valve should be left open for a further 18 min.

Example 6

A ship is 140 m long and displaces 10 000 tonnes is floating at draughts 6.5 m forward and 7.7 m aft. The vessel is scheduled to enter a canal where the maximum draught allowed is 7.2 m. Calculate the minimum amount of cargo to discharge from a compartment which is 30 m forward of the aft perpendicular. The ship's TPC = 16, moment to change trim (MCTC) = 180 and the centre of floatation is amidships.

Change in draught aft $= 7.7 - 7.2 = 0.5\text{ m (50 cm)}$

Change in draught aft $=$ Rise $+$ Change due to Change of Trim

Therefore 50 $$= \frac{w}{TPC} + \frac{1}{2} \times \frac{(w \times d)}{MCTC}$$

$$= \frac{w}{16} + \frac{1}{2} \times \frac{w \times (70 - 30)}{180}$$

$$= \frac{360w + (16 \times 40w)}{16 \times 360}$$

$50 \times 16 \times 360 = 369w + (16 \times 40w)$

$288\,000 = 360w + 640w$

$$w = \frac{288\,000}{1000}$$

w = 288 tonnes represents the minimum cargo to discharge.

Example 7

What do you understand by the term 'loadicator' and what information would you obtain from it?

Answer

A loadicator is the term given to a cargo-loading computer, which is configured to suit the ship's loading programmes. The loadicator will provide the Cargo Officer with the following information once the weight distribution is entered into the programme. Distribution of weights or cargo units in the ship's compartments, the status of relevant tank weights and commodities, the sea-going shear force and bending moment conditions, and the stability aspect with values for measured distance between the keel and the centre of gravity (C of G) of the vessel (KG) and metacentric height (GM). Ballast distribution and quantity would also be available.

The loadicator is often linked to a shoreside monitor to allow data transmission on unit weights for cargo distribution and special stowage requirements. Particularly relevant to a Roll-on, Roll-off (Ro-Ro) vessel engaged on fast turn round, short voyage trades.

Example 8

How would you load a bulk carrier with iron ore?

Answer

Ensure that the hold is clean and that bilge suctions are tested to satisfaction prior to commencing loading. Draw up a pre-load plan and a ballast/deballast plan calculating the stress factors affecting the ship throughout the proposed loading programme. The maximum angle of heel would also be calculated for a potential shift in the cargo volume, bearing in mind that a moisture content is present in the cargo.

The loading rates for the cargo would commence slowly and gradually increase. Fast rates of loading can cause serious damage by generating rapid stress values throughout the ship's length. The important aspect is that iron ore is a dense cargo and heavy. The cargo compartments would only be about one-fourth full. The Chief Officer would calculate the stability based on the load draughts. Condition formats for the bending moment and shear force affecting the loaded condition would be drawn up (stowage factor (SF) iron ore 0.34/0.50).

Example 9

What are the concerns for the Master of a Container Vessel, carrying containers stacked on deck, engaged on the North Atlantic trade in winter?

Answer

The Master, and his Chief Officer would be concerned about the positive stability of the vessel, bearing in mind that the possibility of encountering sub-freezing air temperatures on this trade route at this time is likely. Such conditions could lead to ice accretion, and added weight from icing of the container stack could detrimentally affect the stability of the vessel.

Masters would monitor all weather reports and consider re-routing farther south to warmer latitudes if practical. A reduction in speed could also effectively reduce the rate of ice accretion occurring on the vessel. Where possible, the crew should be ordered to make their best endeavours to remove ice formations if safe to do so.

Example 10

When working as a Cargo Officer aboard an oil tanker, how would you keep the tanks outside the 'flammable limit'?

Answer
The introduction of inert gas into any tank containing hydrocarbon gas/air mixture will decrease the flammable range until a point is reached where the lower flammable limit (LFL) and the upper flammable limit (UFL) coincide. This point corresponds to the oxygen content approximately 11% at which no hydrocarbon gas/air mixture can burn.

> ***Note:*** *Additional reference should be made to the 'Flammability Composition Diagram' found in International Safety Guide for Oil Tankers and Terminals (ISGOTT).*

Example 11
What and when is 'lateral drag' evident and what can the Cargo Officer do to reduce the effects?
Answer
Lateral drag is associated with heavy lifts causing the vessel to heel over as the weight is taken up by the ship's derrick/crane. It can occur during loading or discharging of the load and is effectively a sideways movement of the load as the vessel returns to the upright. If unprepared for, the lateral movement of the load can be violent as the ship rolls against the angle of list.

The effects of lateral drag can be reduced by retaining the line of plumb of the derrick head above the point of landing. This can be achieved by 'coming back' on the topping lift and cargo hoist runner, quickly. This action tends to reduce movement of the load when discharging. If loading the weight a steady slow lifting operation should be carried out.

Example 12
When about to make a heavy lift by means of the ship's heavy derrick, how can the vessel's stability condition be improved so that positive stability is retained throughout the loading period?
Answer
The concern with loading a heavy weight is that the C of G of the weight effectively acts from the head of the derrick. The GM of the ship should be increased by filling the double bottom tanks before the lift is made. This will increase the GM value. Additionally, eliminate any free surface moments in tanks, as this also will reduce the GM value.

Example 13
How can the risk of a grain cargo shifting be reduced?
Answer
Grain should be loaded in accord with the 'Grain Regulations' and the risk of 'shifting' of the cargo can be reduced by:

1. fitting of temporary longitudinal subdivisions (shifting boards)
2. use of bagged cargo in a saucer formation
3. bundling in bulk.

Example 14

How would you describe the 'SF' of a commodity?

Answer

The SF can be defined as that volume that is occupied by a unit weight of cargo and is usually expressed in cubic metres per tonne (m^3/tonne).

$$\text{By example: SF} = \frac{\text{Volume of space}}{\text{Tonnage}}$$

For example, how much cotton at a SF of 2.0 m^3/tonnes could be loaded into a tween deck space of 200 m^3

$$\text{Tonnage} = \frac{\text{Volume of space}}{\text{SF}} = \frac{200}{2}$$
$$= 100 \text{ tonnes cotton.}$$

Example 15

When loading drop trailers and mobile units aboard a Ro-Ro ferry, explain why it is essential that the vessel is kept in the upright position.

Answer

Ro-Ro ferries load their mobile units via vehicle ramps either at the bow or more often through the stern door. These ramps are lowered onto link spans that provide the landing connection between ship and shore. If the vessel develops a list the ramps become angled to the flat shore connection and prevents the movement of vehicles to and from the ship's garage spaces.

Most modern ferries will have automatic stabilizing tank systems to counter any overload to Port or Starboard, so keeping the vessel always in the upright position and vehicle ramps flush on the shore or the link span.

Note: *Over-reliability on tank stabilizers should be avoided and safe practice is always to load and discharge in an even manner to avoid any one side ever become adversely affected by localized tonnage.*

Example 16

When would it be considered appropriate to carry out a 'draught survey'?

Answer

The purpose of a draught survey being conducted would usually be to:

1. ascertain any bending along the length of the vessel, usually after the loading of a bulk cargo
2. determine the exact displacement in order to calculate the total weight of cargo loaded.

Example 17

A vessel is scheduled to load sacks of mail. How would these be loaded on a general cargo vessel if they are loose and not in a container, and what precautions would a prudent Chief Officer take? (Figure B.1)

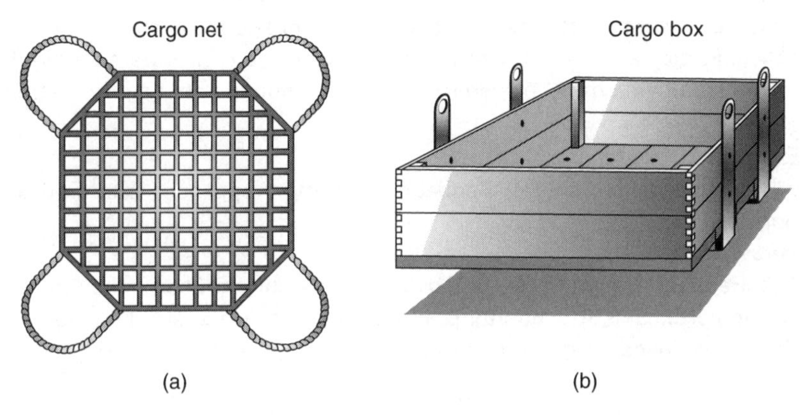

Fig. B.1

Answer

Mails are classed as a specialized cargo and as such would be given lock-up stow.

The bags would be tallied in and tallied out at the ports of loading and discharge, respectively. Watchmen or responsible Ship's Officers would monitor the movement of the mails probably being loaded by means of cargo nets or cargo boxes.

Example 18

What cargo information would the Master of a 'Bulk Carrier' pass to the loading terminal when expecting to berth, to take a full cargo of coal?

Answer

1. In addition to passing the ship's particulars, a pre-loading plan of cargo stowage by hatch, together with the hatch loading order and respective quantities on each pour, assuming that the vessel has sufficient information to prepare such a plan. Confirmation that holds were in a state of readiness to load.
2. The provisional arrival and departure draughts together with details of the ship's own cargo-handling gear and respective capacities of same; details of the ballast capacity and the time required to de-ballast.
3. Additional ship-keeping details reflecting the 'gangway' position, the number of moorings, etc. would also be included as standard information.

Example 19

For what purpose would a Cargo Officer use the 'load density plan'?

Answer

The Ship's Chief Officer would use the load density plan to check the capacity of cargo compartments to ascertain the volume of the space and consider the 'point loading' factor to ensure that the deck strength is adequate to accommodate the intended cargo tonnage to be stowed in the space. Particularly useful with heavy lifts where a concentrated weight over a small area may be seen to exceed the tonnage per square metre.

Example 20

What is contained in the Register of Lifting Appliances and Cargo-Handling Gear?

Answer

The 'Register' is kept up-to-date by the Ship's Chief Officer and contains all the certificates for the lifting appliances, the wires, shackles hooks, chains, etc. used aboard the vessel, for cargo operations.

Appendix C

Codes and conventions affecting cargo work operations, additional references and bibliography

Codes and conventions effecting cargo work operations

Code of Safe Working Practice for Merchant Seaman

Code of Safe Working Practice for the Loading and Unloading of Bulk Cargoes

Inert Gas Code

IMDG, Code (Hazardous Cargoes)

IMO (BLU Code)

IMO, Code of Safe Working Practice for Cargo Stowage

IMO, Construction and Equipment Code (Ships Carrying Dangerous Chemicals in Bulk)

IMO, Grain Regulations

IMO, ISPS Code

International Bulk Cargo Code
International Code for the Safe Carriage of Grain
International Safety Management Code

MARPOL Convention and Subsequent Amendments

Merchant Shipping Regulations for Control of Noxious Liquid Substances in Bulk

Merchant Shipping (Load Lines) (Deck Cargo) Regulations

SOLAS '74 Convention (and Subsequent Amendments)

ICS Tanker Safety Guide (Chemicals)

Additional references

Statutory instruments

S.I. 1509: 1997, The Merchant Shipping (Cargo Ship Construction) Regulations 1997

S.I. 1644: 1999, The Merchant Shipping (Additional Safety Measures for Bulk Carriers)

S.I. 336: 1999 Merchant Shipping-Safety. M.S. (Carriage of Cargoes) Regulations

S.I. 929: 2004, Gas Carrier Amendment Regulations

S.I. 930: 2004, Dangerous or Noxious Liquid Substances in Bulk: Amendment Regulations

MGN 144, The Merchant Shipping (Additional Safety Measures for Bulk Carriers) Regulations 1999

MGN 157, Safety of Personnel during Container Securing Operations, MCA

MIN 154 (M) Safe Loading of Bulk Carriers

Bibliography

Recommendations on the Safe Transport of Dangerous Cargoes and Related Activities in Port Areas, IMO

Cargo Access Equipment, published by Clarke Chapman

Cargo Stowage and Securing, A Guide to Good Practice, Charles Bliault

Code of Safe Practice for Cargo Stowage and Securing, IMO

Code of Safe Practice for Ships Carrying Timber Deck Cargoes, IMO

Design and Operation of Ships Derrick Rigs, British Standards Institute

Hatch Cover Inspections, W. Vervloesem, Nautical Institute

International Code for the Safe Carriage of Packaged Irradiated Nuclear Fuel, Plutonium and High-Level Radioactive Wastes on Board Ships, IMO

International Safety Guide for Oil Tankers and Terminals (ISGOTT), Witherbys

Lashing and Securing of Deck Cargoes, Capt. J.R. Knott, BA.FNI, Nautical Institute

Thomas' Stowage, Brown Son and Ferguson

Commodity and detail index

This commodity index, lists numerous cargo type and product varieties, but it is not exhaustive. Not all items are detailed within the main text and users are advised to reference shipper's documentation as well as other associated works.

Limited detail on commodities is enclosed together with the stowage factor where appropriate and the relevant page number if considered within the main body of this work.

	Commodity name and details	*Stowage factor (m^3/tonne)*	*Page reference*
Acetone	In drums, see IMDG code	2.35–2.5	
Acids	Highly corrosive. IMDG code. Handling precautions required	–	272
Agricultural machinery	In crates	1.39–2.23	
Ammonia	LPG cargo carriage. IMDG code reference	–	174
Ammunition	Magazine stowage. (Dynamite) Dangerous goods IMDG code	Various	261, 265
Anaesthetics	May need temperature controls. Special lock-up stowage as drugs	–	163
Anchor Cable	Flaked flat athwartships in holds and generally overstowed	Variable with cable diameter	
Apples	Temperature control carriage. Cartons	2.37–2.65	198
Apricots	Dried fruit	1.39–1.45	87
	Fresh	2.56–2.78	
Asbestos in cases	See IMDG code	1.53–1.67	
Asphalt	Contains drying oils and liable to spontaneous combustion. Should be properly dry	1.39	
Bacon	Cool stowage	1.73–1.84	198
Bale goods	Various, e.g. Cotton -700 lbs per bale	3.62–3.76	80

(*contd.*)

	Commodity name and details	*Stowage factor (m³/tonne)*	*Page reference*
Bagged goods	Various commodities	2.0–2.23	25, 26, 78–80
Bananas	Mostly in cartons at a carriage temperature of 12/13°C		199–200
	Cartons	3.63–3.90	
Barbed wire	In reels	1.56–1.67	
Barley	Grain regulations. Bulk	1.36–1.50	107
	Bagged	1.45–1.67	
Barrels	Stowed on side bung up, e.g.	1.73–1.78	82, 83
	Castor Oil (also in drums)	1.67–1.73	
	Codfish (salted)	1.90–2.00	
	Colza Oil (also in drums)	1.67–1.73	
	Creosote (also in drums or bulk)	1.67–1.87	161
	Fish Oil (also in bulk)	1.62–167	
	Glucose	1.28–1.34	
	Tung Oil	1.78–1.81	
Bean cargoes	In bags, e.g.		79
	Cocoa, coffee, soya, etc. average S/F	2.17	
	Some products also shipped in bulk		
Beef	(Chilled) cartons	1.53–3.76	197
	(Frozen)	2.37–2.79	
	(Chilled) boneless	1.67–1.74	
Beer	Bottled in cartons or in casks (empty casks/drums)	1.39–1.84	82
	Cartons	1.50–1.56	
	Casks	1.95–2.09	
	Bottled	8.36–9.75	
Bitumen	Inflammable shipped - In solid or liquid form. Will taint other cargoes. (RD 1.00-1.10)		
	Barrels	1.25–1.35	161
	Drums	1.28–1.39	
	Casks	1.53	
Bone meal	Stow clear of edible goods. Bags	1.11–1.25	78
	Bones in bulk	2.23	
Bricks	In crates	0.70	
Bulk cargoes	Various commodities	Varies between 0.31–2.81	100–125
Bullion, bank notes, stamps, etc.	Valuable cargoes Tally in and tally out, lock up stow	Various	71, 88
Butane	LPG cargo carriage		173
Butter	Dairy product in cases	1.45-1.50	197
Cable	Stowed on reels which must be secured against movement	Size variant	88
Canned fruits	Cartons	1.67	

	Commodity name and details	*Stowage factor (m^3/tonne)*	*Page reference*
Carbon Black	Bags often on pallets. Very dirty cargo may sift	1.67	
	Protect other cargoes, see IMDG code		
Cars	designated car carrier. Individual cars may also be shipped in containers or in open stow. 400 mm required between car units	4.18–8.3	223–229
	Crated motor cycles	3.34	
Carpets	Valuable cargo in bales	2.79–3.34	80
Case goods	Various	Size variant	85
Casks	Various commodities, e.g.		86
	China Clay	1.23–1.34	
	Chutney	1.06–1.11	
	Copper Sulphate (highly corrosive)	1.23–1.28	
	Glue (liquid)	3.07–3.34	
	Ginger	1.58–1.81	
	Stearine (natural fat)	1.73–1.78	
Cattle meal cake	Bags or bulk	1.95–2.09	78
Caustic Soda	IMDG code reference. Drums	0.95	
Cement	Different specific gravities. Unitized	0.72–0.79	78, 240
	Bags	0.65–0.70	
	Drums	0.98–1.11	
	Bulk	0.61–0.64	
Cheese	Temperature sensitive		198
	Crates	1.48–1.62	
	Cartons	1.00–1.34	
	Cases	1.20–1.25	
Chemicals	Various		78, 161
	IMDG code, bulk chemical code	–	
China ware/ porcelain	Various packages	3.34–5.57	
	Careful handling - usual for container shipment		
Cinnamon	Highly scented stow away from other cargoes		91
	Bundles	3.62–3.90	
	Cases	2.79	
Cloves	May damage by moisture. Ventilate well and stow away all other goods. Liable to damage		91
	Chests	3.07–3.21	
	Bales	3.07–3.34	
	Bags	3.38–3.42	
Coal	IMDG code	1.25–1.35	91, 117
	Bulk requires surface ventilation		
	Stowage factor variants depending on Country of origin		

(*contd.*)

	Commodity name and details	*Stowage factor (m^3/tonne)*	*Page reference*
Cocoa	Beans in bags	2.0–2.15	79
Coconut oil	Bulk, deep tank carriage	1.06	92
Coffee	Beans in bags	1.81–2.09	79
Coir	(Coconut fibre) in bales	2.79	
Coke	Bulk. Absorbs 20% of its weight in moisture if carried as deck cargo	1.95–2.79	117
Concentrates Bulk	May need shifting boards	Varies on commodity	101, 121
	Average	0.50–0.56	
	e.g. Copper concentrates	0.39–0.50	
	Zinc concentrates	0.56–0.61	
Condensed milk	Cases	1.25–1.28	
Confectionary	In cases or cartons	2.34	
Containers	Generally, goods stowed in containers are under the same conditions as open stow		230–258
Copper	Ingots, ore, coils or concentrates		117-119
	Coils	0.84	
Copra	Stow away from edible foods		79, 122
	Highly infested with copra 'bugs'		
	Troublesome to humans, bulk (hold)	1.95	
	Avoid steelwork contact (tween deck)	2.09–2.15 T/D	
	Bags (hold)	2.09, 2.37 T/D	
Corn	Grain regulations apply, bulk	1.25–1.41	107
	Or in bags	1.39–1.53	
Cotton	Waste. Liable to spontaneous combustion shipped in bales. Cotton goods in cartons. NB. Cotton seed classed as grain, under IMO	3.90–4.46	80
Crude oil	Tanker cargo	(S.G. 0.8/0.9)	126–149
Dairy products	Various. Usually shipped in cartons or cases e.g. Eggs, butter, cheese, etc.	varies with commodity	195, 197, 198
Diesel oil	(S.G. 0.6 / 0.9)		130, 161
Dried Blood	In bags	1.11–1.67	79
Dyes	May be powder, liquid or in paste form. See IMDG code. May cause staining	Varies on package type	86
Earthenware	Mixed parcels		
	Pipes	1.48–1.67	
	Crates	2.79–3.34	
	Cases	1.81–1.95	
	Unpacked	5.57	
Eggs	In boxes	2.93–3.48	198
	In liquid form. Frozen cases and packs	1.11–1.25	

	Commodity name and details	*Stowage factor (m^3/tonne)*	*Page reference*
Elephants	On deck. Full grown animals weigh upto 3 tonnes. Allow for 120 litres of water and 280 kg of food per day. Bills of laden should be endorsed to show that the ship is not responsible for mortality during passage (see livestock)		200, 201
Esparto Grass	(Fibre in bales)	3.62–4.74	81
	Liable to spontaneous combustion		
Ethyl Acetate	Inflammable liquid, drums	1.50–1.78	
Ethyl Chloride	Inflammable liquid, drums	3.62	
Ethylene	Fully refrigerated		174
Explosives	Dangerous Goods. Ammunition, dynamite and fireworks. See IMDG code may require magazine stowage depending on type		261, 265
Fertilizers	In bags or bulk	1.39–1.67	
Fibres	In bales	2.79–3.34	81
Fish	(Frozen)		198
	Boxes or cartons -18° to -15°C	2.50	
	Little danger of taint.		
	Shellfish, crates/cartons	2.28	
	Crustaceans, crates/cartons	2.34	
Fishmeal	Liable to spontaneous combustion		79
	Bags must be well dunnaged to provide adequate ventilation	1.73–1.81	
	Bulk fishmeal may be in pellet or powder form	1.34	
	Space must be full to avoid shifting		
Fish oil	May be shipped in bulk or tins in cases		
	Bulk	1.09	
	Cased tins	1.39–1.48	
Flour	Bags. Keep off steelwork	1.39–1.59	79, 91
Formic acid	Corrosive. Reference IMDG code		
Fruit	Green - clean spaces with mechanical ventilation (extractor fans) Cases or cartons	2.37–2.65	87, 88
	Dried, cases	1.95–2.09	87
	Cartons	1.42	
Fuel oil	(RD 0.92 / 0.99)		161
Furniture	Large packing cases	1.1–2.2	
Garlic in bags	Strong smelling	2.65	87
Gases	(Compressed) in approved cylinders		181, 182

(contd.)

	Commodity name and details	*Stowage factor (m³/tonne)*	*Page reference*
Kerosene	Cases	1.39–1.45	147, 162
	Drums	1.73-1.78	
Lamb	Carcases	4.18	197
	Chilled or frozen carriage - cartons	1.81	
Lard	Liable to melt with heat		
	In cases or pails	1.53–1.61	
	Oil in drums	1.67–1.78	
Latex	Bulk - deep tank stow	1.03	94, 161
	Or drums	1.38–1.53	
Leather	Rolls or bales may be valuable		
	Bales	1.95–2.79	
	Rolls	5.57	
Logs	Different wood types stow at various stowage factors because of differing material densities		
	Teak	2.23–2.37	190–194
	Mahogany sq. logs	0.75–0.84	
Lubricating oil	Usually in cases or 50 gallon drums. Drums may be deck stowed. (RD 0.85 / 0.95)	1.48–1.62	161, 162
Machinery	Sometimes cased	1.12–1.53	
Mail	Lock up stow	2.79–4.18	301
	Parcels	3.34	
Maize	Grain Regulations apply. Bulk	1.25–1.41	107
	Bags	1.39–1.53	
Meats	Chilled of frozen. Cases/cartons	1.81–2.23	197
	Mutton- frozen	2.92–3.06	
	Meat meal in bags	2.23–2.37	
Melons in crates	Adequate ventilation	2.79–3.34	
Molasses	(RD 1.20 / 1.45) Bulk	0.74	94, 162
	Drums	1.39–1.67	
Nitrates	In either bags or bulk. IMDG code reference	1.11	121
Nuts	In bags or bulk. Cool, dry stowage. S/F varies on type	1.95	122
Oakum	In bales	2.51–2.79	81
	Pressed bales	1.95–2.09	
Oats	Liable to heat. Grain Regulations apply. Bulk	1.67–1.94	107
	Bags	1.81–2.06	
Offal	Frozen	2.32–2.37	197

(contd.)

	Commodity name and details	Stowage factor (m^3/tonne)	Page reference
Rice	In bags. Liable to heat and sweat and susceptible to strong odours. Must be kept dry	79, 80, 107	
	Paddy rice	1.81–1.95	
	White rice	1.39–1.45	
Rope	In coils	2.23–2.78	
Rubber	In block or crepe form. Cases	1.90–1.95	81
	Bales	1.81–1.87	
	Sheet	1.67	
	Crepe	3.34	
Rum	See spirits		
Rye	Grain regulations apply. Requires extensive trimming		
	Bulk	1.39	107
	Bags	1.53	
Salt	In bags or bulk. Bags	1.06–1.11	122
	Bulk	0.98–1.11	
Salt rock	Granules. Angle of repose 30°. Bulk	0.98–1.06	
Sand	In bulk	0.53–0.56	
Sanitary ware	In cases/crates	4.18	
Seeds	Stowage factor varies with product	1.28 to 3.76	107
Sheep dip	In drums	1.25–1.53	86
Soda Ash	In bags (Treat as dirty cargo)	1.11–1.25	79
Soya bean	Bulk or bags. Bulk	1.23–1.28	
	Bags, from US	1.59–1.62	
Spirits	In cartons. Inflammable. Special lock up stow, highly pilferable. Also carried in bulk tank containers	1.67–1.81	82
Steel work	Heavy cargo		74–77, 117–119, 272
	Bars	0.33–0.45	
	Billets	0.28–0.39	
	Castings	1.12–1.39	
	Plates	0.28–0.33	
	Pig Iron	0.28–0.33	
Scrap		Various	104, 118
Steel coils	May weigh up to 20 tonnes. Also as pipes, castings and plant machinery	Various	33, 76, 77, 119
Sugar	Dry sugar and Green, (raw wet sugar)		79, 121, 132, 240
	Dry sugar in bulk	1.11–1.25	
	Dry sugar in bags	1.28–1.34	
	Green sugar in bags	1.11–1.17	
Sulphur-	Shipped in bulk. Fire, dangerous cargo, IMDG reference	0.84–0.89	121, 122

(contd.)

The reader should note that modernization, especially the use of containers has rendered many packaging systems obsolete. However, produce packed into containers tends to generally follow the normal standards regarded as necessary for the safe carriage of commodities as 'General Cargo' procedures dictated.

General index